RESPONSIBLE AND EXPLAINABLE ARTIFICIAL INTELLIGENCE IN HEALTHCARE

RESPONSIBLE AND EXPLAINABLE ARTIFICIAL INTELLIGENCE IN HEALTHCARE

ETHICS AND TRANSPARENCY AT THE INTERSECTION

Edited by

AKANSHA SINGH

School of CSET, Bennett University, Greater Noida, Uttar Pradesh, India

KRISHNA KANT SINGH

Delhi Technical Campus, Greater Noida, Uttar Pradesh, India

IVAN IZONIN

Lviv Polytechnic National University, Department of Artificial Intelligence, Lviv, Ukraine

ELSEVIER

ACADEMIC PRESS

An imprint of Elsevier

Academic Press is an imprint of Elsevier
125 London Wall, London EC2Y 5AS, United Kingdom
525 B Street, Suite 1650, San Diego, CA 92101, United States
50 Hampshire Street, 5th Floor, Cambridge, MA 02139, United States

ISBN: 978-0-443-24788-0

For information on all Academic Press publications visit our website at https://www.elsevier.com/books-and-journals

Publisher: Stacy Masucci
Acquisitions Editor: Linda Buschman
Editorial Project Manager: Billie Jean Fernandez
Production Project Manager: Anitha Sivaraj
Cover Designer: Vicky Pearson Esser

Typeset by TNQ Technologies

Contents

Contributors *ix*

1. **Revolutionizing healthcare: The transformative role of artificial intelligence** **1**

Priyansh Praveen Garg, J. Jayashree and J. Vijayashree

Intoduction 1
Artificial Intelligence 2
Healthcare 6
International trends 13
Stages of artificial intelligence in medical procedures 15
AI's potential and advantages in healthcare 16
Requirement of experts 17
Issues and challenges in the adoption of AI in healthcare 18
Reasonableness and imbalance 19
Conclusion 21
References 21
Further reading 23

2. **Ethical considerations in AI powered diagnosis and treatment** **25**

Rashmi Kumari, Subhranil Das and Raghwendra Kishore Singh

Introduction 25
Methodology 28
Ethical issues 29
Strategies for solving ethical issues 43
Applications of AI in the health domain 45
Conclusion 48
References 49

3. **Explainable AI methods to increase trustworthiness in healthcare** **55**

Dmytro Savchuk and Anastasiya Doroshenko

Introduction 55
Methodology 58
Data analysis and visualization 60

Comparison of AI and XAI methods for classification task in healthcare 73
Model comparison 78
Conclusions 85
Acknowledgments 86
References 86

4. Designing transparent and accountable AI systems for healthcare 91

Tripti Sharma, Anuradha Dhull, Akansha Singh and Krishna Kant Singh

Introduction 91
Navigating the ethical landscape: A review of transparent and accountable AI systems in healthcare 95
Frameworks for designing transparent AI systems 100
Conclusion 104
References 105

5. Ensuring fairness and mitigating bias in healthcare AI systems 107

Akansha Singh and Krishna Kant Singh

Introduction to fairness in healthcare AI 107
Understanding bias in AI 113
Sources of bias in healthcare AI systems 114
Fairness in AI 120
Conclusion 123
References 124
Further reading 125

6. AI enhanced healthcare: Opportunities, challenges, ethical considerations, and future risk 127

Soma Gupta, Satarupa Mohanty, Dayal Kumar Behera and Soumya Ranjan Nayak

Introduction 127
Related work 128
AI in healthcare 130
Applying AI in healthcare 135
Ethical principles in AI-powered diagnosis and treatment 140
Challenges of AI in healthcare 148
Future of AI in health care 151

Conclusion 152
References 152

7. **Healthcare revolution: Advances in AI-driven medical imaging and diagnosis** **155**

Amrit Suman, Preetam Suman, Sasmita Padhy, Naween Kumar and Akansha Singh

Introduction 155
Artificial intelligence and medical visualization 158
Smart devices 169
Ambient assisted living 172
References 180
Further reading 181

8. **A deep learning approach for medical image classification using XAI and convolutional neural networks** **183**

Lesia Mochurad, Andrii Ilkiv and Yulianna Mochurad

Introduction 183
State-of-the-arts 185
Materials and methods 191
Modeling and results 200
Analyzing the behavior of classification models when applying XAI methods 203
Conclusions 217
Funding 217
Acknowledgments 217
References 218

9. **Hybrid ensemble learning model to improve the performance and interpretability of medical diagnosis: Small data tasks** **221**

Ivan Izonin, Roman Tkachenko, Veronika Kachmar and Myroslav Havryliuk

Introduction 221
Background 223
Hybrid cascade-based prediction model and its algorithmic implementations 229
Datasets description and their preprocessing 235
Modeling and results 239
Comparison and discussion 247

Conclusions 255
Funding 256
References 256

10. Legal and regulatory issues related to AI in healthcare 259

Anshita Gera, Anuradha Dhull, Akansha Singh and Krishna Kant Singh

Introduction 259
AI in healthcare: A gift or a threat? 263
Mainstream use cases involving AI's legal and regulatory concerns in healthcare 266
Conclusion 280
References 281
Further reading 283

11. Responsible and explainable artificial intelligence in healthcare: Conclusion and future directions 285

Akansha Singh, Krishna Kant Singh and Ivan Izonin

Introduction 285
Major challenges and resolutions 291
Future directions 293
Conclusion: Shaping the future of AI in healthcare 295
References 296
Further reading 297

Index 299

Contributors

Dayal Kumar Behera
School of Computer Engineering, KIIT Deemed to Be University, Bhubaneswar, Odisha, India

Subhranil Das
School of Computer Science, University of Petroleum and Energy Studies, Dehradun, Uttarkhand, India

Anuradha Dhull
Department of CSE, The NorthCap University, Gurugram, Haryana, India

Anastasiya Doroshenko
Lviv Polytechnic National University, Institute of Computer Science and Information Technologies, Department of Automated Control Systems, Lviv, Ukraine

Priyansh Praveen Garg
School of Computer Science and Engineering, Vellore Institute of Technology, Vellore, Tamil Nadu, India

Anshita Gera
Department of CSE, The NorthCap University, Gurugram, Haryana, India

Soma Gupta
School of Computer Engineering, KIIT Deemed to Be University, Bhubaneswar, Odisha, India

Myroslav Havryliuk
Lviv Polytechnic National University, Department of Artificial Intelligence, Lviv, Ukraine

Andrii Ilkiv
Lviv Polytechnic National University, Department of Artificial Intelligence, Lviv, Ukraine

Ivan Izonin
Lviv Polytechnic National University, Department of Artificial Intelligence, Lviv, Ukraine

J. Jayashree
School of Computer Science and Engineering, Vellore Institute of Technology, Vellore, Tamil Nadu, India

Veronika Kachmar
Lviv Polytechnic National University, Department of Artificial Intelligence, Lviv, Ukraine

Naween Kumar
School of CSET, Bennett University, Greater Noida, Uttar Pradesh, India

Rashmi Kumari
School of Computer Science Engineering and Technology, Bennett University, Greater Noida, Uttar Pradesh, India

Lesia Mochurad
Lviv Polytechnic National University, Department of Artificial Intelligence, Lviv, Ukraine

Yulianna Mochurad
Danylo Halytsky Lviv National Medical University, Lviv, Ukraine

Satarupa Mohanty
School of Computer Engineering, KIIT Deemed to Be University, Bhubaneswar, Odisha, India

Soumya Ranjan Nayak
School of Computer Engineering, KIIT Deemed to Be University, Bhubaneswar, Odisha, India

Sasmita Padhy
School of Computing Science and Engineering, VIT Bhopal University, Bhopal, Madhya Pradesh, India

Dmytro Savchuk
Lviv Polytechnic National University, Institute of Computer Science and Information Technologies, Department of Automated Control Systems, Lviv, Ukraine

Tripti Sharma
Department of CSE, The NorthCap University, Gurugram, Haryana, India

Akansha Singh
School of CSET, Bennett University, Greater Noida, Uttar Pradesh, India

Krishna Kant Singh
Delhi Technical Campus, Greater Noida, Uttar Pradesh, India

Raghwendra Kishore Singh
School of CS & AI, SR University, Warangal, Telangana, India

Amrit Suman
Sharda School of Engineering & Technology, Sharda University, Greater Noida, Uttar Pradesh, India

Preetam Suman
School of Computing Science and Engineering, VIT Bhopal University, Bhopal, Madhya Pradesh, India

Roman Tkachenko
Lviv Polytechnic National University, Department of Publishing Information Technologies, Lviv, Ukraine

J. Vijayashree
School of Computer Science and Engineering, Vellore Institute of Technology, Vellore, Tamil Nadu, India

CHAPTER 1

Revolutionizing healthcare: The transformative role of artificial intelligence

Priyansh Praveen Garg, J. Jayashree and J. Vijayashree
School of Computer Science and Engineering, Vellore Institute of Technology, Vellore, Tamil Nadu, India

Intoduction

The protection and improvement of people's physical, mental, and social well-being is the noble goal of healthcare, which has long been at the forefront of human development. The world of healthcare has always placed a great deal of reliance on the profound understanding and experience of its human practitioners. The complex art of diagnosis, treatment, and compassionate care forms its roots. Nevertheless, the unrelenting march of technology, driven by the unrelenting advancement of AI, has spurred a revolutionary change in the way healthcare is delivered that has the potential to upend the entire structure of the sector.

The empathic touch of the practitioner, the intuition developed through years of practice, and the unshakable dedication to patient welfare have historically been the beating hearts of healthcare. The advent of AI, however, adds a potent ally to this equation: intelligent computer systems painstakingly built to mimic and enhance human cognitive powers. These AI systems have the potential to profoundly alter how we prevent, diagnose, treat, and manage diseases. This paradigm shift in healthcare will go far beyond gains in productivity and efficiency.

Nowadays, AI refers to any task that a machine can carry out at least as well as a human. AI represents the ambitions and worries of a sector seeking more intelligent solutions. The study of artificial intelligence crosses several disciplinary boundaries, including computer science, psychology,

Responsible and Explainable Artificial Intelligence in Healthcare
ISBN 978-0-443-24788-0
https://doi.org/10.1016/B978-0-443-24788-0.00001-7

linguistics, philosophy, and neuroscience. Reasoning, information, planning, learning, natural language processing, perception, and the capacity to move and manipulate objects are among the primary (Artificial intelligence (AI), n.d.).

Furthermore, AI's promise goes far beyond the confines of the current healthcare environment. It permeates the intricate workings of EHRs, provides answers to the age-old problem of antibiotic resistance, creates ground-breaking brain-computer interfaces, and opens up the field of immunotherapy for the treatment of individualized cancers (Ramesh et al., 2004).

AI has achieved mastery in medical imaging (Ramesh et al., 2004), a field where it evaluates complex pictures such as X-rays, MRIs, and CT scans with unmatched accuracy. With their unerring eye, these deep learning algorithms let radiologists spot even the most minute irregularities, enabling quicker diagnosis as well as the early detection of diseases, which ultimately improves patient outcomes.

In conclusion, the combination of artificial intelligence with healthcare is more than just a technological one; it is a promise. A promise to redefine the full scope of patient care while also enhancing human potential. Even though there are still challenges, they are only rest breaks on this evolving route. With further research, innovation, and a dedication to the ethical and responsible integration of AI, the future of healthcare may be brighter, more compassionate, and more precise than it has ever been. The advancement of healthcare into a realm of unfathomable possibilities, where the pursuit of wellbeing reaches new heights and transforms not only the industry but also the people it touches, is what human knowledge and artificial intelligence collaborate to achieve in the future.

Artificial Intelligence

The concept of AI marks a significant milestone in the history of technology and computer science. It was John McCarthy who first coined this term in 1956 during a conference that served as a pivotal moment in the formal recognition and establishment of AI as a field of study. AI is a multifaceted domain within computer science that is primarily concerned with the development and design of intelligent computer systems capable of emulating and simulating human intelligence. What sets AI apart is its exceptional capacity to process natural language, learn from data, and strategize actions, effectively enabling machines to perform tasks that would traditionally demand human cognitive abilities (International Journal of Engineering Research, n.d.).

The ambitious objective of emulating human cognitive abilities is at the heart of AI's purpose. It aims to develop electronic things that behave like human specialists in a variety of fields, including healthcare, acting as independent, autonomous systems. AI has changed and grown throughout the years, playing a more prominent role in our daily lives and assuming various functions in a variety of fields. Personal assistants such as Siri and Alexa, automated mass transit, aviation technology, improved computer games, facial recognition for passport checks, voice-activated virtual assistants, the development of driverless cars, and even companion robots that assist and communicate with humans are examples of these manifestations.

AI technologies have become potent allies for medical professionals in the field of healthcare. They are excellent at assessing health data, providing insightful commentary, and helping doctors make knowledgeable decisions about patient treatment. This capacity for in-depth analysis of health data improves our comprehension of patients' potential future healthcare requirements, encouraging more proactive and individualized medical interventions. Notably, the incredible adaptability and versatility of AI technology are two of its distinguishing features. It can easily integrate with current technologies to improve their functionality. In disciplines such as chemistry and medicine, where AI-enhanced computers may start and support common diagnostic procedures, this adaptability is particularly clear. AI helps minimize time-consuming chores and administrative hassles by streamlining these procedures, thus increasing the effectiveness of healthcare delivery.

AI is interdisciplinary and draws on a wide range of disciplines, including computer science, engineering, machine learning, chemistry, biology, physics, astronomy, neuroscience, and the social sciences. AI does not work in a vacuum. This transdisciplinary nature is evidence of the numerous uses of AI across a range of societal domains.

Understanding that AI is not a single, monolithic technology but rather a rich tapestry of computational models and algorithms is vital. Numerous specialized fields congregate within the AI landscape, each bringing distinct skills to the larger AI ecosystem. Among these important disciplines are (International Journal of Engineering Research, n.d.; AlphaFold, n.d).

Expert systems

Expert systems are computer programs created to mimic the judgment skills of human decision-makers in particular fields. To make wise decisions, they rely on a knowledge base of facts and regulations.

Applications: Expert systems are used in many fields, including engineering (machine problem diagnostics), finance (investment advice), and healthcare (diagnosis and recommended treatments).

Relevance: By codifying the knowledge of human experts, they provide a methodical way to tackle complicated problems and make this information available for decision assistance.

Fuzzy logic

Fuzzy logic is used to reason and make decisions when faced with uncertainty. Fuzzy logic allows for degrees of truth, in contrast to traditional binary logic, enabling more complex decision-making.

Fuzzy logic is used in consumer electronics such as washing machines, robotics such as vacuum cleaners, and control systems such as vehicle cruise control to manage uncertain input and make judgments.

Meaning: It makes AI systems more flexible to real-world circumstances by enabling them to operate with hazy or unclear information.

Artificial neural networks

ANNs are computer models that draw inspiration from the structure and operation of the human brain. They are employed for tasks such as pattern recognition and data analysis and are composed of interconnected nodes (neurons) arranged in layers.

Applications: image and speech recognition, natural language processing, and recommendation systems (such as Netflix movie suggestions) all make extensive use of ANNs.

Relevance: ANNs are useful for tasks involving data analysis and classification because they are excellent at learning intricate patterns from huge datasets.

Machine learning

Definition: The development of methods and models that allow machines to learn from data and enhance their performance on a particular activity without being explicitly programmed is the broad field of machine learning.

Applications: Machine learning is pervasive and employed in many fields, such as fraud detection, autonomous driving, personalized marketing, and predictive analytics.

Meaning: By enabling machines to learn from and adapt to new data, it automates decision-making processes and produces more precise forecasts and insights.

Natural language processing

The goal of NLP is to make it possible for computers to comprehend, analyze, and produce human language. It involves activities such as sentiment analysis, language translation, and text analysis.

Applications include text summarization tools, language translation services, chatbots, and virtual assistants such as Siri and Alexa.

Importance: NLP closes the communication gap between humans and robots, allowing for efficient conversation and linguistic comprehension (Alverez-Valle and Moore, n.d.).

Computer vision

Machines are now able to interpret and comprehend visual data from the outside world, thanks to computer vision. It involves activities including facial recognition, picture segmentation, and object detection.

Applications: Computer vision is used in augmented reality (such as Snapchat filters), surveillance systems, driverless vehicles, and medical picture analysis.

Importance: It makes up a crucial component of AI systems that interact with or alter photos and videos because it enables robots to "see" and comprehend the visual world.

Robotics

Robotics is the integration of AI with mechanical devices (robots) to enable autonomous or partially autonomous task performance. Applications for robotics are numerous, ranging from industrial and healthcare to space exploration.

Applications for robotics include autonomous drones for diverse uses, surgical robots for minimally invasive surgery, and industrial robots for manufacturing.

Meaning: Robotics brings AI into the physical world by enabling machines to interact with it and carry out mobile and manipulative tasks (Bajwa et al., 2021).

Deep learning

Deep learning is a subset of machine learning that utilizes deep neural networks, which include several layers. When it comes to tasks such as speech and picture recognition, these networks are incredibly effective.

Applications: The success of applications such as facial recognition, autonomous driving, language translation (such as Google Translate), and recommendation systems is attributed to deep learning.

Meaning: Deep learning has made strides in jobs requiring high-level abstractions and feature extraction from unstructured data, making it a key player in the development of AI (Bajwa et al., 2021). Each of these subfields adds to AI's wide range of applications, which are constantly increasing and changing the way we interact with technology and live our lives.

In conclusion, a new era of opportunities and capacities has begun as a result of the development of AI. Its effects go far beyond technology; they affect the entire nature of human progress and the possibility of a more effective, knowledgeable, and customized future. While obstacles still exist, the combined efforts of researchers, scientists, and engineers are pushing the limits of what AI is capable of, changing the world in ways that were previously only conceivable in science fiction (Dickson, n.d.).

Healthcare

The maintenance and improvement of people's physical, mental, and social well-being is a multidimensional area of healthcare. It includes a range of tasks, such as the management of diseases, illnesses, injuries, and disabilities, as well as their prevention, diagnosis, and treatment. In the past, healthcare has relied heavily on human judgment and decision-making. However, the way healthcare is provided and administered has undergone a profound transformation since the emergence of AI.

Artificial intelligence, or AI, is the term used to describe the creation of intelligent computer systems that can imitate human cognitive processes. AI is a useful technology that has the potential to transform a number of areas of the healthcare sector. Using the data provided, the following examples show how AI is useful in healthcare (AlphaFold, n.d; Alverez-Valle and Moore, n.d.; Bejnordi et al., 2017).

- In order to interpret the results of medical imaging with outstanding precision, deep learning techniques are used by AI in radiology. AI's capacity to spot minute changes in images that humans would miss can be useful to radiologists. This increases accuracy while also quickening the diagnostic procedure (Ramesh et al., 2004).
- Oncology/Cancer: AI algorithms have proven to be more effective than human doctors at identifying cancer, notably lung and breast cancer. They are better than humans at complicated dataset analysis, enabling earlier identification and more individualized treatment strategies.
- Telemedicine: By transferring medical data and images across healthcare facilities, AI-enabled telemedicine makes it easier to diagnose and treat patients remotely. This helps to increase access to healthcare and is especially useful when providing care in rural or underserved areas (Bibel and Petkov, 1984).
- Electronic Health Records: AI software can examine EHRs, which provide detailed patient histories. By analyzing this data, AI can forecast how diseases will develop, enabling early intervention and customized treatment regimens.
- Mobile Health (mHealth): Powered by AI, mobile health makes use of portable electronics such as smartphones and wearables to enhance the delivery of healthcare. These gadgets are transformed into health management platforms by AI algorithms and data analytics, improving access to healthcare services, the effectiveness of treatments, and possibilities for preventive care (Davies, 2019).
- Medical Research: AI supports medical research by evaluating big and complex datasets, spotting patterns, and looking up pertinent publications in the scientific literature. Additionally, it can forecast negative drug reactions, predict illness outbreaks, and match suitable people with clinical studies (Dickson, n.d.).

The impact of AI in healthcare is significant. It improves healthcare services' accuracy, effectiveness, and accessibility. Large volumes of data can be swiftly and correctly analyzed, enabling earlier diagnosis, more efficient treatments, and better patient outcomes. Furthermore, AI-driven healthcare solutions can aid in bridging access gaps to care, particularly in rural or disadvantaged locations. The potential uses of AI in healthcare are many, and they hold out the prospect of a time when technology and human expertise will collaborate to offer people and communities the best possible care. **AI applications in healthcare today and in the future**.

The future of the healthcare business is bright because of the expanding use of AI. In the short, medium, and long terms, the following information covers the current situation and potential applications of AI in healthcare.

Artificial intelligence today

Unlike human doctors, AI systems in healthcare are not yet reasoning machines. They are not capable of using "common sense" or "clinical intuition and experience." As a signal translator, AI is more effectively employed to translate patterns from datasets. Its main function is to automate laborious, frequent, and repetitive processes (Esteva et al., 2019).

Healthcare businesses are starting to employ AI systems today, especially in areas where they excel at processing massive volumes of data quickly. AI has previously shown effectiveness in precision diagnostics, such as in the detection of diabetic retinopathy and the support of radiotherapy planning, for instance. These applications mark the start of artificial intelligence's influence in healthcare (Raumviboonsuk et al., 2019; Sadiku et al., 2018).

Medium-term AI

The creation of more potent AI algorithms is predicted to make significant progress in the medium term. Less data will be needed to effectively train these algorithms, making them more efficient. Given the amount of unstructured healthcare data, they will also be able to use unlabeled data, which is a desirable skill. Additionally, AI systems will become proficient at merging a variety of data types, including unstructured data such as medical photos, behavioral data, and pharmaceutical data, as well as structured data such as electronic health records.

It is anticipated that healthcare organizations and medical practices will advance from being passive adopters of AI platforms to becoming active coinnovators in collaboration with digital firms. The creation of cutting-edge AI systems specifically designed for precision therapies will result from this joint approach. AI will become more significant in improving decision-making and treatment planning.

Long-term

Long-term predictions indicate that AI systems will grow to be even smarter and more powerful. Precision medicine will be attained by AI-enhanced healthcare systems, radically altering the healthcare environment. The traditional one-size-fits-all approach to medicine will give way to a more proactive, individualized, and data-driven method of managing disease as a result of this revolution (Esteva et al., 2019).

Key characteristics of this proposed healthcare system include.

- Prevention: AI will make it possible to identify health hazards early and put preventative measures in place that are specific to each patient's profile.
- Personalization: Treatment strategies and treatments will take a patient's particular genetic, behavioral, and clinical information into account.
- Data-driven: To improve patient outcomes and both the patient and clinical experiences, the entire healthcare ecosystem will heavily rely on data.
- Cost-Effective: A more cost-effective delivery system is anticipated to result from the implementation of AI-driven, preventative, and personalized healthcare approaches.

Overall, the long-term vision for AI in healthcare is to enhance patient and clinical experiences, improve patient outcomes, and accomplish these objectives within a more effective and efficient healthcare delivery system. The healthcare sector is anticipated to have substantial and wide-ranging repercussions as a result of this paradigm change, which will eventually benefit both patients and healthcare professionals (Raumviboonsuk et al., 2019).

Connected/augmented care

AI has the potential to significantly reduce inefficiency in the healthcare industry, improve patient flow and experience, and enhance caregiver experience and patient safety throughout the care pathway. For instance, AI could be applied to the remote monitoring of patients (for example, intelligent telehealth through wearables and sensors) in order to identify and provide timely care to patients who are at risk of deterioration.

We anticipate that in the not-too-distant future, healthcare clinics, hospitals, social care services, patients, and caregivers will all be connected to a unified, interoperable digital infrastructure employing passive sensors in conjunction with ambient intelligence. This will allow for seamless data sharing across all of these entities. The following are two examples of how AI is being used in linked healthcare.

AI-powered chatbots and digital personal assistants

Patients are using AI chatbots to diagnose symptoms and recommend future actions in community and primary care settings. It is possible for AI chatbots to be linked with wearable devices such as smartwatches in order to provide patients and caregivers with insights that can help them improve their behavior, sleep, and overall wellness.

Ambient and intelligent care

We also take note of the emergence of ambient sensing that does not require the use of any peripherals, which is a wireless, touchless sensor and machine learning platform for remote monitoring of sleep, breathing, and behavior that was founded by faculty and researchers at the Massachusetts Institute of Technology.

The Google Nest platform boasts the ability to track and analyze sleep patterns, even those that are disrupted by activities such as coughing (Gulshan et al., 2016).

A paper that was just recently published investigates whether or not it is possible to use smart speakers to monitor cardiac rhythms in a contactless manner (Gulshan et al., 2016).

Automation and ambient clinical intelligence: AI systems that leverage natural language processing technology have the potential to automate administrative tasks such as documenting patient visits in electronic health records, optimizing clinical workflow, and enabling clinicians to focus more time on caring for patients.

Precision diagnostics

Diagnostic imaging

The most advanced application of AI now available is one that allows for the automated classification of medical images. A recent analysis of AI and ML-based medical devices that have been certified for usage in the United States of America and Europe between the years 2015 and 2020 discovered that more than half of these devices (129 (58%) devices in the United States and 126 (53%) devices in Europe) have been approved or CE-marked for radiological applications. Studies have shown that artificial intelligence is capable of meeting or exceeding the performance of human experts in image-based diagnoses from several medical specialties, such as pneumonia in radiology (wherein a convolutional neural network trained with labeled frontal chest X-ray images outperformed radiologists in detecting pneumonia), dermatology (wherein a convolutional neural network was trained with clinical images and was found to classify skin lesions accurately), and pathology (wherein one study trained a convolutional neural network with clinical images (Jiang et al., 2017; Kassam and Kassam, 2020).

We are aware that the NHS already has some successful projects in this field (such as the University of Leeds Virtual Pathology Project and the National Pathology Imaging Co-operative), and we anticipate that AI-based diagnostic imaging will see widespread adoption and expansion in the medium future (Khurana et al., 2022). Here are two examples of how such technologies could be put to use.

Diabetic retinopathy screening

Screening individuals for the early detection and treatment of diabetic retinopathy is essential to achieving the global goal of lowering the vision loss that can be prevented as a result of diabetes. However, due to the large number of diabetic patients and the limited number of eye care professionals available across the globe, screening can be quite expensive. Research conducted in the United States, Singapore, Thailand, and India on the use of automated AI algorithms for the diagnosis of diabetic retinopathy has shown reliable diagnostic performance and cost effectiveness (Mabiyan, 2018; Mahajan and Kaur, 2022; Mason, 2003). In addition, the Centers for Medicare and Medicaid Services granted Medicare coverage for the use of the artificial intelligence algorithm known as "IDx-DR," which was approved by the Food and Drug Administration and had a sensitivity of 87% and a specificity of 90% when it came to detecting diabetic retinopathy that was more severe than mild (Raumviboonsuk et al., 2019; Sadiku et al., 2018).

Improving the precision and reducing waiting timings for radiotherapy planning

The use of AI to assist physicians with picture preparation and planning activities for radiation cancer treatment is an important use of the technology.

At the moment, segmentation of the images is a tedious and time-consuming job that is carried out manually by an oncologist using specialized software to create contours around the regions of interest. The open-source technology called InnerEye, which is based on artificial intelligence, can minimize this preparation time for head and neck cancer as well as prostate cancer by up to 90%, which means that the amount of time spent waiting to begin potentially life-saving radiotherapy treatment can be drastically reduced (Mintz and Brodie, 2019).

Precision therapeutics

In order to make headway towards the development of precision therapies, our knowledge of disease needs to undergo significant expansion. Researchers all over the world are investigating the cellular and molecular foundation of disease, accumulating a wide variety of multimodal datasets that have the potential to lead to digital and biological biomarkers for the diagnosis, severity, and progression of the disease.

Both immunomics and synthetic biology, as well as drug development, are important future applications of artificial intelligence.

Immunomics and synthetic biology

It is possible that in the future, through the application of AI tools on multimodal datasets, we will be able to better understand the cellular basis of disease as well as the clustering of diseases and patient populations in order to provide more targeted preventive strategies. One such strategy would be the use of immunomics to diagnose and more accurately predict care and treatment options. This will usher in a new era of revolutionary change across numerous care standards, with a particular influence in the areas of oncology, neurology, and rare diseases. It will also personalize the experience of care for each unique patient.

AI-driven drug discovery

Clinical trial design will be much improved by AI, medication manufacturing processes will be optimized, and generally, any combinatorial optimization process in healthcare could be substituted by AI. With the recent announcements by DeepMind and AlphaFold, we have already seen the beginnings of this, which now sets the stage for greater comprehension of disease processes, protein structure prediction, and the development of more focused therapies (for both rare and more prevalent diseases) (Mondal et al., 2023; Muoio, n.d.).

Precision medicine

New curative therapies

Over the past decade, synthetic biology has produced developments such as CRISPR gene editing and some personalized cancer therapies. However, the life cycle for developing such advanced therapies is still extremely inefficient and expensive. In the future, with better access to data (genomic, proteomic, glycomic, metabolomic and bioinformatic), AI will allow us to handle far more systematic complexity and, in turn, help us transform the way we understand, discover and affect biology. This will improve the efficiency of the drug discovery process by helping to better predict early which agents are more likely to be effective and also better anticipate adverse drug effects, which have often thwarted the further development of otherwise effective drugs at a costly late stage in the development process. This, in turn, will democratize access to novel, advanced therapies at a lower cost.

AI-empowered healthcare professionals

In the longer term, healthcare professionals will leverage AI to augment the care that they provide, which will allow them to provide safer, more effective, and more standardized care at the top of their license. For instance, clinicians could use an "AI digital consult" to examine "digital twin" models of their patients (a truly "digital and biomedical" version of a patient), which would enable them to "test" the effectiveness, safety, and experience of an intervention (such as a cancer drug) (Esteva et al., 2019).

International trends

The field of medicine is beginning to see the application of artificial intelligence. As AI technology gets closer to globalization, it is becoming increasingly important to monitor both the actions taken by various governments and the changes made to regulatory frameworks all over the world. There have been governmental shifts and investments made all around the world in AI. Artificial intelligence has emerged as a top strategic priority for governments all over the world. The following are some instances of common patterns observed on a global scale (Muoio, n.d.).

Canada: Canada has a one-of-a-kind chance, albeit a time-limited one, to become world leaders in the design of systems that make use of AI technology. It has solidified its position as the preeminent authority in the field of research into AI-related technologies. Its primary cities, Montreal, Toronto, and Vancouver, have become centers for research and development in artificial intelligence, which has attracted major corporations including Google, Facebook, Uber, Microsoft, and Samsung. The development of human-centered artificial intelligence is one of Canada's top priorities, along with fostering worldwide scientific collaboration (Oktay et al., 2020).

The United Kingdom has declared that one of its goals is to position the country as a global leader in artificial intelligence and data technology. As part of the United Kingdom's aim to lead global regulation on AI ethics, the country's government has established the Centre for Data Ethics and Innovation (Centre for Data Ethics and Innovation).

France: The French government has ambitious goals to become a world leader in artificial intelligence development. The following are some of its primary goals and initiatives.

(1) developing an open data policy to encourage the adoption and application of AI in industries such as healthcare;
(2) establishing a regulatory and financial framework to support the development of domestic "AI champions";
(3) putting in place regulations to ensure that AI developments remain transparent, explainable, and nondiscriminatory.

China's government has aspirations to manufacture items that are intelligent and networked, such as automobiles, service robots, and identification systems. It makes an announcement on investments in the industry's training resources, standard testing, and cybersecurity.

India: The "AIforAll" initiative is part of India's overall AI policy, which intends to broaden access to artificial intelligence. Additionally, India is making efforts to position itself as an "AI Garage," which would make the artificial intelligence technology that is developed in India accessible to the rest of the globe (Oktay et al., 2020).

As a result of the high number of patients, India has a wealth of data. The field of healthcare in India is becoming increasingly interested in implementing artificial intelligence and machine learning, respectively. However, India faces a number of challenges, such as an aging population, inadequate infrastructure, restricted access to healthcare facilities, low treatment adherence rates, and a shortage of care professionals.

The accessibility, quality, and effectiveness of global health systems might all be improved with the help of AI technologies. The concentration of worldwide attention on the formulation of open and accountable AI policy is a pattern that can be observed everywhere.

Stages of artificial intelligence in medical procedures

The application of artificial intelligence, also known as AI, in the field of medicine has, without a doubt, led to considerable advancements in terms of both productivity and precision. The categorization of artificial intelligence's involvement in medical care into four stages provides a useful foundation for comprehending the uses of this technology. Let's explore these stages further.

Locating the problematic region that requires intervention

1. Artificial intelligence systems can be used to aid in the early diagnosis and precise localization of anomalies or diseases. For instance, artificial intelligence may examine medical pictures such as X-rays, MRIs, and CT scans in addition to endoscopy in order to detect cancers, fractures, and other diseases.
2. AI-driven robotic surgical systems can make minimally invasive procedures, also known as keyhole surgery, more precise, hence minimizing the amount of patient trauma and the amount of time needed for recovery.

Eliminate the following possibilities

1. Pattern recognition and the examination of patient data are two of the ways that AI might assist in excluding possible diagnoses. It is able to evaluate the patient's symptoms, test results, and medical history in order to propose possible diagnoses and rank the necessary diagnostic tests.
2. The correct diagnosis of complicated diseases is made much easier by the use of machine learning algorithms, which can examine massive databases to find connections and risk factors.

Clinical studies that are both more effective and faster

1. The process of finding new pharmaceuticals and developing them can be sped up with the help of AI, which can predict how possible drugs would interact with biological systems. This speeds up the research process, which in turn leads to clinical trials being completed more quickly.
2. It is possible for machine learning algorithms to select potential candidates for clinical trials based on patient data. This improves the efficiency of recruiting while simultaneously shortening the duration of clinical trials.

Developing diagnostic biomarkers in order to identify diseases

1. Artificial intelligence can be used to help identify biomarkers and genetic markers that are connected with particular diseases. These markers have the potential to assist in both the early diagnosis and the tailored therapy processes.
2. As was said earlier, ultrasound guidance is just one example of how artificial intelligence might improve the precision of medical treatments. It can assist medical professionals target particular tissues or organs with greater accuracy, hence minimizing the level of invasiveness and the amount of patient discomfort.

The use of AI has had a significant impact on the healthcare industry as a whole, with benefits ranging from increased diagnostic accuracy to streamlined processes in drug development and treatment planning. However, it is essential to make certain that these AI applications are developed and deployed in an ethical manner, under appropriate supervision, in order to guarantee the safety of patients and the confidentiality of their data (Quinn et al., 2021).

AI's potential and advantages in healthcare

- Improvements in Quality and Safety: By offering more precise diagnoses and treatment predictions, AI has the ability to greatly improve the Quality and Safety of Patient Care. Better patient health outcomes may result from this.
- Cost reduction: By enhancing diagnostic precision and therapeutic efficacy, AI can aid in lowering medical expenses. As a result, healthcare systems and individuals may spend less on healthcare.
- Productivity Multiplication: AI-powered health aides can provide care for millions of patients at once, significantly boosting accessibility and productivity in the industry.
- AI can reduce the likelihood of human error in medical procedures, improving patient safety.
- treatments are streamlined thanks to AI, which makes expensive and time-consuming medical treatments simpler and more affordable.

New Era of Patient Care: With the incorporation of AI, a new era of patient care is about to begin, with technology taking center stage in the provision of healthcare.

Requirement of experts

The application of artificial intelligence frameworks presents considerable challenges for medical professionals. The debate over whether AI will eventually replace or enhance human doctors is the most important aspect. As is the case in a number of other domains, the presence of even insufficiently automated frameworks results in significant changes to the expert community, as certain positions are eliminated and others are created. Existing frameworks for medical applications may be generally considered to be examples of "feeble AI." This means that they can succeed at "limited" tasks and be useful in assisting human doctors in dismal, tedious activities such as locating components or sectioning organs or structures in photographs. According to the information provided, clinical claims to fame controlling the visual evaluation of photos have started testing generous modifications using such "guides." The most important thing to take into consideration is that contemporary systems are able to process a significantly larger number of photographs than any human administrator can, without experiencing fatigue or falling behind schedule, and, if properly prepared, with startlingly superior results.

Computer-based intelligence frameworks likewise demonstrate extraordinary execution in the acknowledgment of common language and composed writings, which opens the best approach to separating data—for frameworks to gain from numerous accessible sources, for example, clinical records, which are not normally accessible in organized, valuable configurations.

In any event, when all is said and done, clinical AI frameworks do not have the capacity to "decipher" the unique circumstance and "produce" the most particular human highlights (inventiveness, sentiments), and so they cannot be considered (yet) to be "solid" or "general" AI. Despite the fact that there are still many questions to be answered, AI systems have the potential to improve many aspects of a specialist's work, and those specialists who choose not to use them will become obsolete.

The availability of massive amounts of information that convey many different types of data about each individual patient necessitates the development of new expert profiles that aren't necessarily doctors in order to study and isolate the aspects of the data that are most beneficial for diagnosis and treatment. Hereditary guides are currently becoming a part of therapeutic groups in order to assist not only patients but also professionals in comprehending the bewildering quantity of information related to quality. The proliferation of new information sources, within the foreseen vision of "extended customized medication," will most likely necessitate the development of distinct "clinical information researcher" profiles in order to provide specific direction for both medical professionals and their patients.

Medical care is not just a scholarly connection but also extremely friendly, which is particularly significant in situations of fundamental significance, such as choices with implications for life quality, palliative consideration, and willful elimination. Moreover, medical care is especially vital in situations where it is necessary to choose between life and death.

There are many different points of view regarding trust, compassion, and the evolution of care, and how they can be influenced by the new reality of automated frameworks. These points of view range from the reduction of clinical mistakes to the possibility of AI frameworks assisting people with making difficult decisions to the planning of better specialist AI and patient-AI association strategies. All of these points of view have the potential to be impacted by the new reality of automated frameworks. Defenders of AI-improved human specialists thought new technologies to be a wonderful way to recover human components of the connection between the specialized and the patient; nonetheless, the actual effects are still unknown at this time (Quinn et al., 2021).

Issues and challenges in the adoption of AI in healthcare

- Healthcare industry Complexity: Compared with certain other industries, the adoption of AI has been slower in the healthcare industry due to its complexity.
- Data security and patient privacy must be protected at all costs. AI businesses must manage data protection laws while leveraging priceless medical data.
- Job worries: There is a myth that artificial intelligence would displace medical professionals. Healthcare professionals have distinctive qualities and skills that are difficult to quantitatively mimic, despite the fact that AI is a useful tool.
- Ethical considerations: There are serious ethical considerations around the use of AI in healthcare, including worries regarding the suitability of AI applications.

- Regulatory Gaps: The absence of formal rules governing AI usage in healthcare complicates adoption and raises concerns about accountability and safety.
- Healthcare expenses are rising at unsustainable rates, and by not implementing AI technology, chances for cost savings and life-saving interventions are being hampered.
- Technological Obstacles: Data preparation, knowledge extraction, interpretability, and the demand for openness and responsibility in AI development and application are among the technological challenges facing AI implementation in the healthcare industry.
- Data Access and Quality: Harnessing AI's full potential in healthcare is difficult due to problems with data access and quality.
- Building the requisite organizational capabilities and technical infrastructure for the deployment of AI can be difficult and time-consuming.
- Safety and Regulation: To safeguard patient interests and maintain ethical practices, it is essential to ensure the safe and regulated use of AI in healthcare.

As a result, even while AI has the amazing potential to revolutionize healthcare, its implementation is fraught with several difficulties and issues. To optimize AI's beneficial effects on healthcare systems and patient care, it is crucial to balance its advantages with related ethical, governmental, and technical issues. Realizing AI's potential to alter the healthcare industry will depend in large part on how well these problems are solved (Dickson, n.d.; Esteva et al., 2019).

Reasonableness and imbalance

In the context of artificial intelligence's application in medicine and healthcare, the junction of reasonability and precariousness creates a complicated and ever-evolving terrain. We investigate the consequences of the evolving artificial intelligence market in healthcare, taking into account not just the positive potential of the market but also the ethical and social considerations that arise.

- **Positive Developments in the Application of AI in Healthcare**

  It is expected that the incorporation of AI into medicine and healthcare will result in favorable impacts all around the world. Diagnostics, treatments, and overall patient care could all benefit from the application of AI, which would ultimately lead to better health outcomes.

Analyze financière Artificial intelligence has the potential to cut expenses in certain areas of healthcare, which is important financially. For instance, the price of translating a human genome is on the low end of the spectrum, which makes personalized therapy more approachable.

A huge step forward in terms of the capabilities of healthcare is represented by the field of precision medicine, in which AI interventions make it possible to produce highly individualized medicines that are specifically matched to the genetic make-up of an individual patient.

- **Considerations of an Ethical and Social Nature**

Disparities in Cost: Despite the fact that AI-driven tailored treatments have enormous promise, there may be significant costs associated with using them. There is a possibility that some AI-intervened drugs, particularly tailored treatments, could be exceedingly pricey, which could result in serious access issues to cutting-edge medical care.

Inequality on the Rise The rapid advances in clinically noteworthy medicines powered by AI have the potential to worsen existing inequities in healthcare access and outcomes. This could have a negative impact on patients. Those who are able to pay for these therapies may enjoy significant health benefits, while those who cannot may be left behind in terms of their health.

In order to address the potential for expanding healthcare inequality, it may be necessary to develop new models of health inclusion, protection, and moderation. This could be a requirement for addressing the danger of growing healthcare inequality. These models might have the goal of ensuring that AI-driven healthcare advancements help a greater proportion of the general population.

Thus, artificial intelligence has the potential to change the healthcare industry and enhance patient outcomes. However, it also poses a number of ethical and societal concerns, particularly with regard to cost and accessibility. A significant difficulty that needs to be addressed by healthcare systems, entrepreneurs, and governments is finding a way to strike a balance between maximizing the benefits of AI in medicine and ensuring that everyone has equal access to the latest developments in this field. As AI continues to have an impact on healthcare, there is an increasing need for new models of inclusion and protection, which highlights the necessity of taking proactive ethical and societal factors into account (Rakhra et al., 2021).

Conclusion

In conclusion, the application of AI in healthcare is a fundamental revolution that comes with enormous potential as well as formidable obstacles. Healthcare solutions that are personalized, accurate, predictive, and portable are predicted by AI to be a feature of the future. However, there are many other steps involved in achieving this aim.

Healthcare personnel may be empowered by AI by automating chores, which frees them up to concentrate on what really matters—patient care. It has the ability to raise patient care standards globally, lower medical costs by providing accurate diagnoses and treatments, and increase patient quality and safety.

A digital renaissance in healthcare is being driven by AI, which is at a critical juncture in the fusion of medicine and technology. In addition to digitizing medical records, the next 10 years will focus on mining these digital resources for insightful data. AI will be essential in turning these insights into practical clinical results.

The use of AI in healthcare is not without its difficulties, though. Care must be taken to address concerns about data privacy, ethical issues, legal and ethical frameworks, and the application of AI algorithms in the actual world. To fully utilize the potential of AI, healthcare management must also engage in the digital upskilling of their workforce.

Despite the issues and challenges that surround AI in the healthcare sector, it continues to be a disruptive force that has the power to completely transform the sector. A seamless continuum of care with a focus on early detection and prevention is what the future of healthcare promises, and this is all made possible by AI technologies.

Thus AI's place in healthcare is changing quickly. The combined efforts of healthcare practitioners, researchers, and policymakers offer the possibility of utilizing AI's ability to enhance healthcare on a worldwide scale, despite the enormous difficulties that still need to be solved. It is obvious that AI will continue to influence healthcare in a positive way as we stand at the nexus of medicine and technology.

References

Alverez-Valle, G.J. Moore, Project InnerEye Open-Source Deep Learning Toolkit: Democratizing Medical Imaging AI, n.d.

Artificial Intelligence (AI), Healthcare, n.d.

Bajwa, J., et al., 2021. Artificial intelligence in healthcare: transforming the practice of medicine. Future Healthc. J. 8 (2), e188–e194. https://doi.org/10.7861/fhj.2021-0095.

Bejnordi, B.E., et al., 2017. Diagnostic assessment of deep learning algorithms for detection of lymph node metastases in women with breast cancer. JAMA, J. Am. Med. Assoc. 318 (22), 2199–2210. https://doi.org/10.1001/jama.2017.14585.

Bibel, W., Petkov, B., 1984. Artifical intelligence Methodology, systems, applications. In: Proceedings of the International Conference on Artifical Intelligence: Methodology, Systems, Applications (North-Holland).

Davies, S.E., 2019. Artificial intelligence in global health. Ethics Int. Aff. 33 (2), 181–192. https://doi.org/10.1017/S0892679419000157.

B. Dickson, How Artificial Intelligence Is Revolutionizing the Mhealth Industry, n.d.

Esteva, A., et al., 2019. A guide to deep learning in healthcare. Nat. Med. 25 (1), 24–29. https://doi.org/10.1038/s41591-018-0316-z.

Gulshan, V., et al., 2016. Development and validation of a deep learning algorithm for detection of diabetic retinopathy in retinal fundus photographs. JAMA 316 (22). https://doi.org/10.1001/jama.2016.17216.

Jiang, F., et al., 2017. Artificial intelligence in healthcare: past, present and future. Stroke Vasc. Neurol. 2 (4), 230–243. https://doi.org/10.1136/svn-2017-000101.

Kassam, A., Kassam, N., 2020. Artificial intelligence in healthcare: a Canadian context. Healthc. Manag. Forum 33 (1), 5–9. https://doi.org/10.1177/0840470419874356.

Khurana, R., et al., 2022. Energy-efficient fog-assisted system for monitoring diabetic patients with cardiovascular disease. Blockchain and deep learning for smart healthcare. In: An Ensemble Deep, pp. 323–352.

Mabiyan, R., 2018. How Artificial Intelligence Can Help Transform Indian Healthcare. ETHealthWorld.

Mahajan, R., Kaur, P., 2022. Artificial intelligence in healthcare. Int. J. Health Sci. 4505–4512. https://doi.org/10.53730/ijhs.v6ns2.5987.

Mason, R.O., 2003. Ethical Issues in Artificial Intelligence. Elsevier BV, pp. 239–258. https://doi.org/10.1016/b0-12-227240-4/00064-2.

Mintz, Y., Brodie, R., 2019. Introduction to artificial intelligence in medicine. Minim Invasive Ther. Allied Technol. 28 (2), 73–81. https://doi.org/10.1080/13645706.2019.1575882.

Mondal, S.S., et al., 2023. EDLDR: an ensemble deep learning technique for detection and classification of diabetic retinopathy. Diagnostics 13 (1). https://doi.org/10.3390/diagnostics13010124.

D. Muoio, Google's next-gen Nest Hub debuts with contactless sleep monitoring and analysis features, Mobi Health News, n.d.

Oktay, O., et al., 2020. Evaluation of deep learning to augment image-guided radiotherapy for head and neck and prostate cancers. JAMA Netw. Open 3 (11), E2027426. https://doi.org/10.1001/jamanetworkopen.2020.27426.

Quinn, T.P., et al., 2021. Trust and medical AI: the challenges we face and the expertise needed to overcome them. J. Am. Med. Inf. Assoc. 28 (4), 890–894. https://doi.org/10.1093/jamia/ocaa268.

Rakhra, A., Gupta, R., Singh, A., 2021. Medical Imaging and Diagnosis Using Machine Learning and Deep Learning, pp. 390–394. https://doi.org/10.1201/9781003193838-71.

Ramesh, A.N., et al., 2004. Artificial intelligence in medicine. Ann. R. Coll. Surg. Engl. 86 (5), 334–338. https://doi.org/10.1308/147870804290.

Raumviboonsuk, P., et al., 2019. Deep learning versus human graders for classifying diabetic retinopathy severity in a nationwide screening program. npj Digital Medicine 2 (1). https://doi.org/10.1038/s41746-019-0099-8.

Sadiku, M.N.O., Zhou, Y., Musa, S.M., 2018. Natural language processing in healthcare. Int. J. Adv. Res. Comput. Sci. Software Eng. 8 (5). https://doi.org/10.23956/ijarcsse.v8i5.626.

AlphaFold: A Solution to a 50-Year-Old Grand Challenge in Biology", The AlphaFold team, n.d.

International Journal of Engineering Research and Advanced Technology (IJERAT) E-ISSN", n.d. 10.31695/IJERAT.2020.3670.

Further reading

Garnham, A., 1988. Artifical Intelligence an Introduction.

Sadiku, M.N.O., Shadare, A.E., Musa, S.M., 2017. Int. J. Eng. Res. 6, 450–452.

Sadiku, M.N.O., Ashaolu, T.J., Musa, S.M., 2019a. Artificial intelligence in medicine: a primer. Int. J. Trend Res. Dev. 6, 270–272.

Sadiku, M.N.O., et al., 2019b. A primer, pp. 43–46.

Sadiku, M.N.O., et al., 2019c. Teleeverything Phenomena, pp. 35–38.

Saravanan, V., et al., 2021. Impact of big data in healthcare system-a quick look into electronic health record systems. In: Machine Learning and the Internet of Medical Things in Healthcare. Elsevier, United States, pp. 251–262. https://doi.org/10.1016/B978-0-12-821229-5.00009-4.

Senior, A.W., et al., 2020. Improved protein structure prediction using potentials from deep learning. Nature 577 (7792), 706–710. https://doi.org/10.1038/s41586-019-1923-7.

Singh, A., Kant Singh, K., 2024. Blockchain and deep learning: research challenges, open problems, and future. In: Blockchain and Deep Learning for Smart Healthcare. CRC Press, India, pp. 425–431. https://www.wiley.com/en-se/Blockchain+and+Deep+Learning+for+Smart+Healthcare-p-9781119792383.

Singh, A., et al., 2023. Energy-efficient fog-assisted system for monitoring diabetic patients with cardiovascular disease. In: Blockchain and Deep Learning for Smart Healthcare. Wiley, pp. 323–352. https://doi.org/10.1002/9781119792406.ch13.

Sniecinski, I., Seghatchian, J., 2018. Artificial intelligence: a joint narrative on potential use in pediatric stem and immune cell therapies and regenerative medicine. Transfus. Apher. Sci. 57 (3), 422–424. https://doi.org/10.1016/j.transci.2018.05.004.

Strodthoff, N., Strodthoff, C., 2019. Detecting and interpreting myocardial infarction using fully convolutional neural networks. Physiol. Meas. 40 (1). https://doi.org/10.1088/1361-6579/aaf34d.

Ting, D.S.W., et al., 2019. Artificial intelligence and deep learning in ophthalmology. Br. J. Ophthalmol. 103 (2), 167–175. https://doi.org/10.1136/bjophthalmol-2018-313173.

Wang, X., Peng, Y., Lu, l, 2017. Hospital-scale chest x-ray database and benchmarks on weakly-supervised classification and localization of common thorax diseases. IEEE CVPR 2017, 2097–2106.

Wang, A., et al., 2021. Using smart speakers to contactlessly monitor heart rhythms. Commun. Biol. 4 (1). https://doi.org/10.1038/s42003-021-01824-9.

Xie, Y., et al., 2020. Artificial intelligence for teleophthalmology-based diabetic retinopathy screening in a national programme: an economic analysis modelling study. Lancet Glob. Health 2 (5), e240–e249. https://doi.org/10.1016/S2589-7500(20)30060-1.

CHAPTER 2

Ethical considerations in AI powered diagnosis and treatment

Rashmi Kumari[1], Subhranil Das[2] and Raghwendra Kishore Singh[3]
[1]School of Computer Science Engineering and Technology, Bennett University, Greater Noida, Uttar Pradesh, India; [2]School of Computer Science, University of Petroleum and Energy Studies, Dehradun, Uttarkhand, India; [3]School of CS & AI, SR University, Warangal, Telangana, India

Introduction

Over the past few years, artificial intelligence has experienced significant advancements, exerting a substantial influence on various aspects of people's lives, such as healthcare (Panesar and Panesar, 2021), intelligent transportation (Das et al., 2020, 2021), and education (Beck et al., 1996). In March 2019, as part of several significant reforms, the World Health Organization declared the creation of a new Digital Health department. This global commitment aligns with various national plans unveiled in recent years as governments confront the challenges posed by AI in the healthcare sector. Numerous instances of AI-driven digital health applications are currently accessible, with some having obtained market authorization. Judging by the substantial private investment in digital health, it appears that the future pipeline of digital health products will be abundant. ML, as a fundamental component of AI, has propelled AI development stage (Xu, 2019; Badillo et al., 2020). ML approaches can empowered the machines in order to handle complicated problems without executing algorithms (Xu, 2019). Deep learning (DL) has been applied in Artificial Neural Networks (ANN) and is particularly adept at addressing intricate problems with unstructured data (Kumari et al., 2022b, 2023b). Among various domains, healthcare stands out as the ML applications (Srivastava et al., 2017). The application of AI techniques in healthcare has the potential to expedite cancer detection, enhance medical diagnoses, and facilitate the care and monitoring of the elderly through robotic assistance (Kumari et al., 2022a;

Responsible and Explainable Artificial Intelligence in Healthcare
ISBN 978-0-443-24788-0
https://doi.org/10.1016/B978-0-443-24788-0.00002-9

2023a,b). ML techniques, capable of processing data contribute to accurate measurements and future implications (Esteva et al., 2017). Despite ML's pivotal role in advancing AI, ethical concerns, particularly in the healthcare sector, have been raised (Lee et al., 2017).

Recognizing ethics as a key focus in advancing and implementing AI across various sectors (Gibney, 2020; Ouchchy et al., 2020). The ethical challenges that arise have implications for the ongoing progress and adoption of AI, particularly in healthcare. In this domain, technology must adhere to legal frameworks, regulations, and privacy principles to safeguard collective welfare (Esteva et al., 2017). Examining ethical challenges associated with AI in biomedical research, patient care, and public health, the focus shifts to a systemic oversight model for governing AI innovation in the health sector (Beck et al., 1996). The subsequent discussion explores potential approaches to tackle emerging ethical challenges in this rapidly evolving domain. The overall objective is to establish the groundwork for an ethically responsible development of AI in health research and clinical practice domains.

Within the landscape of artificial intelligence, the integration of sophisticated machine learning (ML) algorithms, employing Deep Learning (DL) and other intricate techniques, often results in the creation of black-box models. As indicated in recent research (Carabantes, 2020), these models exhibit low transparency and explainability characteristics, presenting a challenge to developers and end-users. In the development process, even those who create these AI systems find it challenging to articulate the decision-making processes embedded within black-box models. This lack of internal clarity extends to users, who are confronted with decisions devoid of accompanying explanations (Stahl et al., 2022). Concerns are raised by the ML algorithms' intrinsic "black-box" nature, especially in high-stakes situations where decisions can have dire repercussions if they are not fully understood or explained. The healthcare sector, marked by its critical implications for human lives, exemplifies the complexities associated with the adoption of AI. In this domain, the integration of AI faces various barriers, including issues of accuracy, explainability, transparency, and compatibility. These factors collectively contribute to a reluctance to embrace AI technologies in everyday medical practice fully. Recognizing the importance of explainability in AI algorithms, recent discourse underscores its role in fostering responsible AI practices. Beyond technical considerations, explainability becomes a pivotal element in building trust and encouraging user engagement with AI systems (Shin, 2021). This emphasis on transparency and clarity in algorithmic decision-making has garnered attention from diverse stakeholders, from governmental bodies and the media to the legal community (Hofman et al., 2017). In response to these concerns, the research community has actively embraced the challenge of designing AI systems with improved explainability. Various efforts have been directed toward creating models that not only deliver results but also offer insights into the decision-making processes (Zeng et al., 2017; Weller, 2019). Despite these advancements in explainability, the integration of AI-enabled solutions in healthcare remains entangled with multiple ethical concerns, gradually emerging as decisive factors influencing the broader adoption of AI in the field. This dynamic landscape reflects an ongoing dialogue between technological innovation and the imperative to address ethical considerations, particularly in domains where human well-being is at stake.

In response to the ethical risks, policymakers as well as professionals have sought various approaches. Notable examples include the European Commission's (Buruk et al., 2020; Dixit et al., 2021), and the Chinese government's "Beijing AI Principles" (Li et al., 2010). From these entities, the European Union (EU) helps in establishing an ethical framework for AI (Madiega, 2019). Unlike other guidelines, the EU emphasizes a "human-centered" approach, aligning with European values (Buruk et al., 2020). The EU framework addresses ethical issues that are relevant to global settings; these issues are globally meaningful, based on fundamental rights. These guidelines, rooted in a fundamental rights perspective, contribute to the framework supporting the majority of the UN Sustainable Development Goals (SDGs) (Bærøe et al., 2020). Notably, these guidelines span all industrial sectors and do not explicitly focus on the ethical and legal aspects of AI in healthcare.

Certainly, there are scholarly publications delving into the ethical dimensions of AI in healthcare. Notable works include Morley's research group's "The ethics of AI in health care" (Morley et al., 2020), Gerke's group's examination (Gerke et al., 2020), and Reddy's group's proposition (Reddy et al., 2020). Morley's group stands out for its approach to systematically mapping ethical issues within AI in healthcare, classifying them under epistemic, normative, and overarching perspectives (Morley et al., 2020). On the other hand, Gerke's group emphasizes the ethical considerations intertwined with legal challenges in AI-driven healthcare but does not conduct a comprehensive review of how AI can impact these challenges (Gerke et al., 2020). Lastly, Reddy's group contributes to the

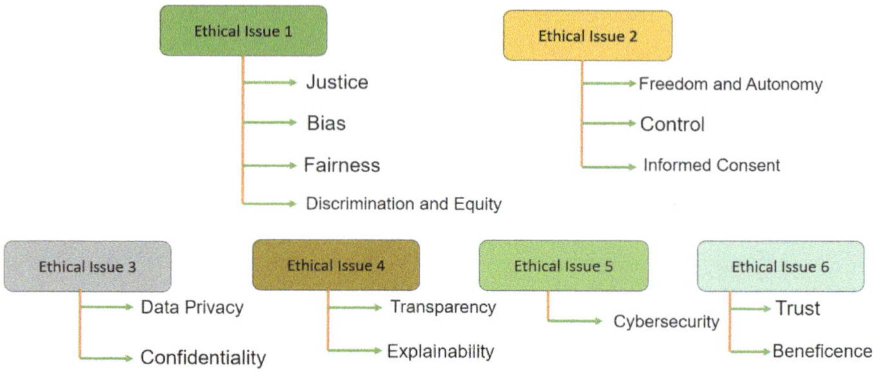

Figure 2.1 Ethical issues in AI systems.

discourse by introducing and implementing a governance model (Reddy et al., 2020). In essence, these publications collectively offer diverse perspectives on the ethical intricacies associated with the integration of AI in healthcare, ranging from a detailed mapping of ethical concerns to considerations of legal challenges and the proposal of a governance framework for effective AI implementation in the healthcare sector. ethical issues are shown in Fig. 2.1.

In this comprehensive literature review, our primary goal was to outline the recognized ethical considerations associated with the integration of AI as follows:

1. A delineation of the ethical issues has been explored in the services related to healthcare.
2. An exploration of the ethical strategies is explained where the intention was to equip development teams engaged in AI for adopting proactive measures in addressing and managing ethical concerns throughout the design processes of their AI systems.

Methodology

To conduct our study, we employed a systematic literature review, a method that entails the comprehensive evaluation and interpretation of all pertinent primary studies related to a specific research question. A number of methodological steps were used in this study, including developing the strategy for searching the literature, establishing eligibility requirements, carrying out the selection process, conducting quality evaluations, and synthesizing the data that was collected. Notably, the research process involved the active participation of three independent reviewers. In our pursuit of comprehensive insights, we meticulously conducted a literature search across five prominent databases such as, IEEE, ACM, PubMed, Nature-SCI, and others. Distinguished by their academic prominence, the first three databases were complemented by the inclusion of the AI Ethics Guidelines Global Inventory, which serves as a repository for published frameworks articulating ethical values in AI. This methodological choice ensured a well-rounded exploration, encompassing both scholarly research and the nuanced perspectives found in gray literature. The latter, particularly valuable for its wealth of principles and guidelines governing ethical AI, extends beyond the traditional academic realm. Our meticulous selection process targeted documents that addressed key facets, including artificial intelligence, healthcare, ethics, and guidelines. Employing an automated search mechanism in each designated search engine, we strategically deployed keywords and synonyms from Table 2.1.

The Distribution of the number of the different journals published with respect to different parameters are shown in the following figures. Fig. 2.2 shows the Ethical Issue 1, Fig. 2.3 shows the Ethical Issue 2, Fig. 2.4 shows the Ethical Issue 3, Fig. 2.5 shows the Ethical Issue 4, Fig. 2.6 shows the Ethical Issue 5,Fig. 2.7 shows the Ethical Issue six.

Ethical issues

According to the literature studies, it has been categorized as six major ethical issues where the subethical issues, such as justice, bias, and freedom has been explained in subsections. Moreover, the supplementary ethical issues have been explained in the last subsection explaining its importance in AI applications in the healthcare industry.

Ethical issue one

Justice: The equitable and fair distribution of the benefits and burdens these systems provide among various populations is at the center of the ethical question of justice in AI systems. Concerns arise when AI algorithms exhibit biases that disproportionately impact certain demographic groups, leading to systemic inequalities. The lack of representativeness in training data, coupled with inherent biases embedded in algorithms, can result in discriminatory outcomes, reinforcing existing societal disparities. Additionally, issues of transparency and accountability in decision-making processes within AI systems contribute to challenges in ensuring that the

Table 2.1 Overview of the distribution of the ethical issues published in different journals.

Journals	Ethical issue 1	Ethical issue 2	Ethical issue 3	Ethical issue 4	Ethical issue 5	Ethical issue 6	Supplementary ethical issue
IEEE							
SCI/SCIE	13	12	18	8	17	15	22
SCOPUS	6	8	5	3	7	3	8
ESCI	19	14	12	7	19	13	15
Nature							
SCI/SCIE	26	24	14	20	6	7	14
SCOPUS	21	11	13	5	7	9	16
ESCI	12	16	8	4	10	3	10
ACM							
SCI/SCIE	10	12	13	4	5	5	6
SCOPUS	17	10	6	8	4	2	5
ESCI	8	4	6	5	7	9	3
PubMed							
SCI/SCIE	13	15	4	10	14	7	9
SCOPUS	8	6	3	7	5	9	4
ESCI	3	7	9	5	2	8	3
Others							
SCI/SCIE	10	17	15	13	15	18	9
SCOPUS	21	19	16	18	8	9	11
ESCI	7	9	5	12	10	8	14

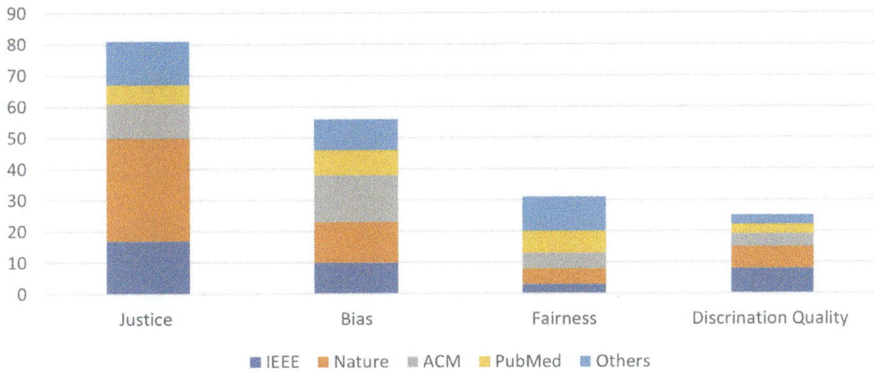

Figure 2.2 Distribution of the number of the different journals published with respect to ethical issue one.

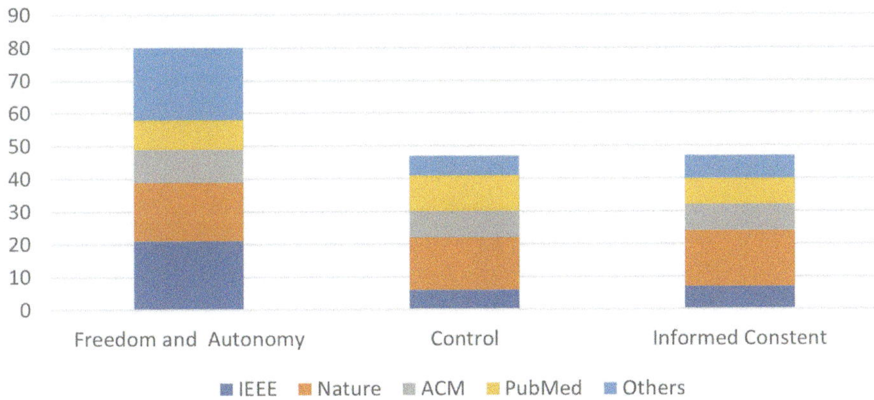

Figure 2.3 Distribution of the number of the different journals published with respect to ethical issue two.

benefits of AI are accessible to all, regardless of socioeconomic status, race, or other demographic factors. Addressing the ethical imperative of justice in AI systems involves not only mitigating biases but also fostering inclusivity, transparency, and fairness in the design, deployment, and impact assessment of these technologies.

Bias: Within the healthcare domain, the genesis of bias predominantly stems from the integration of self-learning algorithms during their learning processes (Artificial intelligence and medical imaging 2018: French radiology community white paper, 2018; Jain et al., 2023). Termed as "algorithmic bias," this phenomenon refers to the systematic errors or

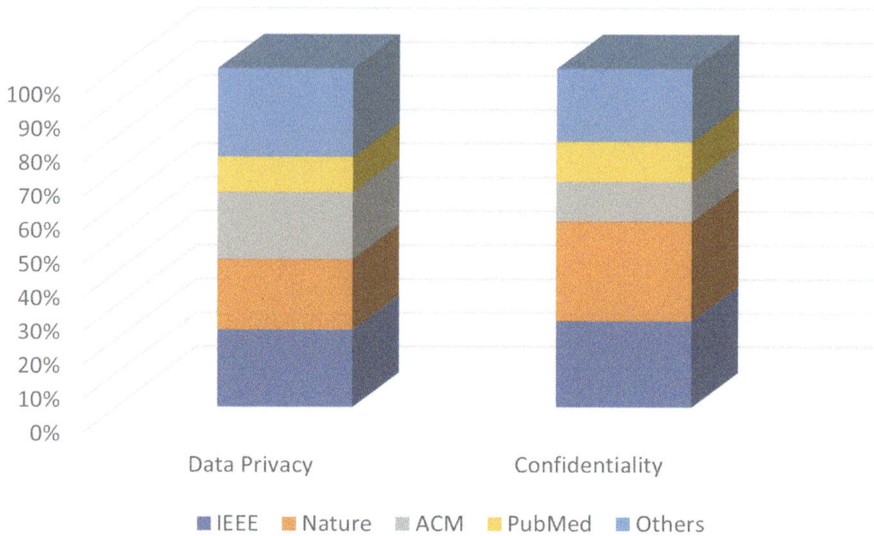

Figure 2.4 Distribution of the number of the different journals published with respect to ethical issue three.

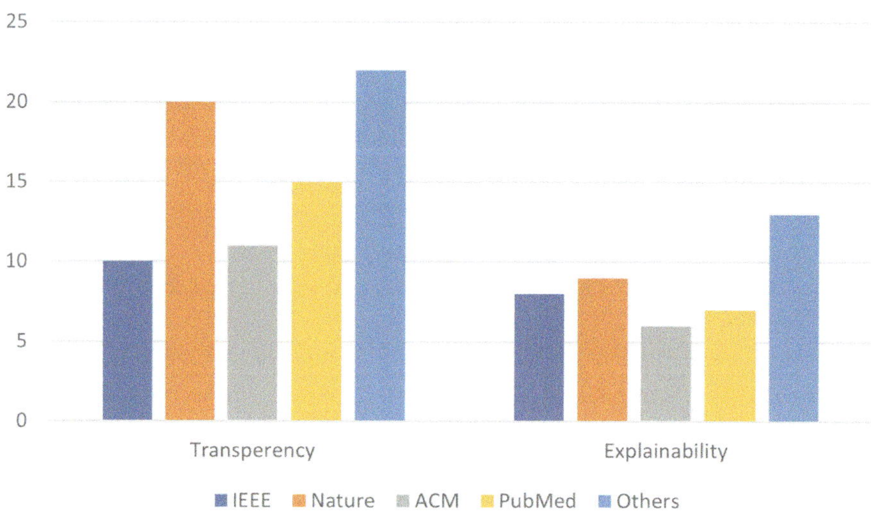

Figure 2.5 Distribution of the number of the different journals published with respect to ethical issue four.

unintended tendencies of AI algorithms, resulting in a preference for one outcome over another (Raymond Geis et al., 2019). Diverse manifestations of algorithmic bias, such as self-fulfilling prophecies, overfitting, and black-box issues, contribute to this complexity (Amann et al., 2020). Additionally,

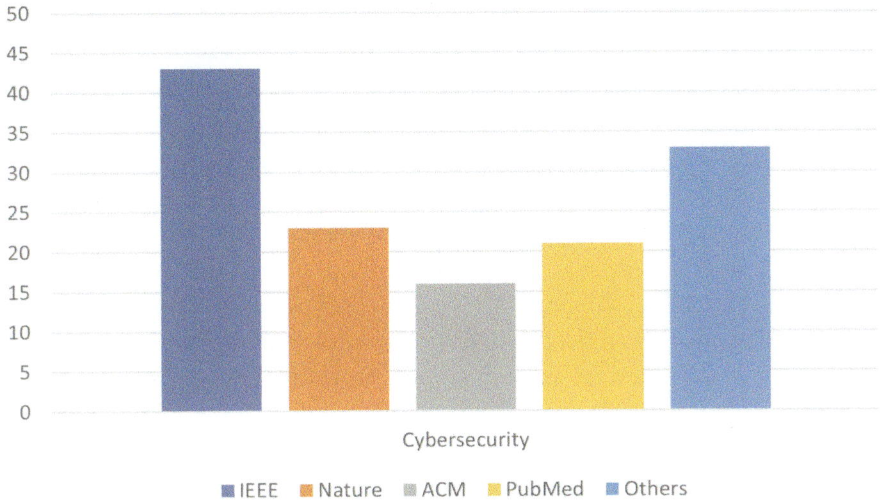

Figure 2.6 Distribution of the number of the different journals published with respect to ethical issue five.

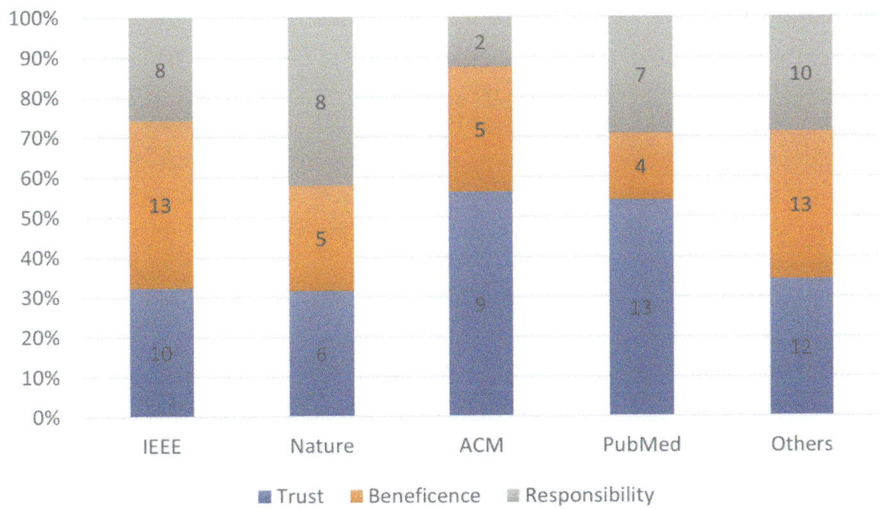

Figure 2.7 Distribution of the number of the different journals published with respect to ethical issue six.

the quality and representativeness of training data used as inputs emerge as pivotal contributors to bias-related challenges. When training data, which is frequently inappropriate or insufficiently representative, does not fully capture the spectrum of the target population, or when the system operates with incomplete knowledge, input biases result. The nuanced interplay of algorithmic intricacies and training data quality underscores the imperative of addressing these biases comprehensively to ensure the ethical and equitable deployment of AI technologies in healthcare.

Fairness: Fairness emerged as the third subissue within the broader ethical concern with 10 selected documents. Rooted in the fundamental principles of justice and equity, fairness has remained a prominent ethical consideration for centuries, particularly alongside the discourse on bias (Ahmad et al., 2020). The intricacies and opacity inherent in most machine learning algorithms significantly impact the fairness of AI systems, resulting in decision-making processes and outputs that are challenging to explain, thereby complicating the assessment of the system's fairness. Notably, the literature distinguishes two primary factors within the domain of fairness: gender and ethnicity. Historical instances, such as the 1970s case involving St George's Hospital, highlight the use of algorithms that discriminated based on gender (Ahmad et al., 2020). Moreover, the decisions made by the organizations have deleted the data based on gender biasedness for legal, institutional, or commercial reasons, may inadvertently compromise fairness, potentially heightening the risk of indirect discrimination.

Discrimination and Equality: The inherent data-driven nature of AI technology lays the groundwork for potential discrimination, primarily rooted in the curation of training datasets. The equality can be compromised if the trained data lacks the representation or if target demographics are inadequately defined. A pertinent example is highlighted by France Assos Santé, underscoring concerns about the potential risk selection practices by insurance companies amidst the computerization of health data (Toulouse et al., 2018). Furthermore, the utilization of aggregated data for decision-making regarding broader populations or the creation of novel demographic groups introduces the possibility of discriminatory outcomes. This emphasizes the critical need for meticulous attention to the selection and composition of training datasets to mitigate biases and uphold the principles of fairness in AI applications. The percentage of the articles related to ethical issue one have been published in different journals has been shown in Fig. 2.8.

Ethical issue two

Freedom and Autonomy: The possible effects on personal liberties and the autonomy of decision-making are at the center of the ethical debate

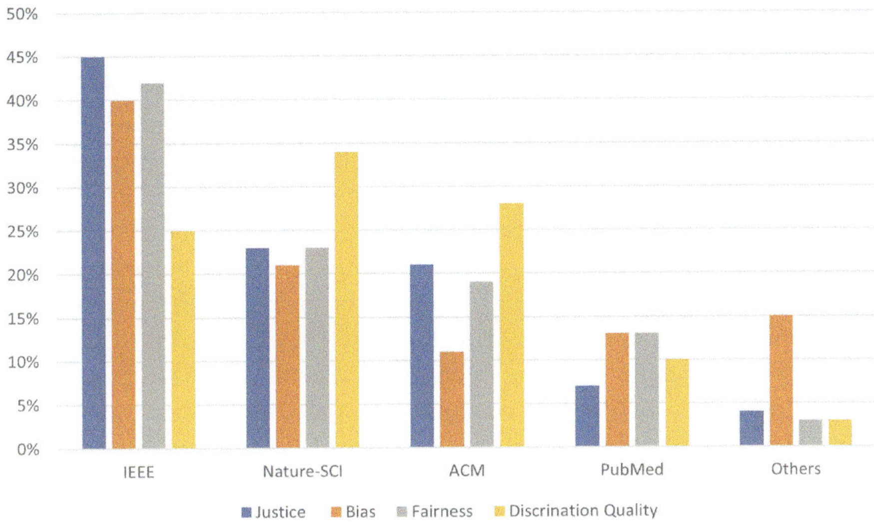

Figure 2.8 Percentage of articles published for ethical issue one.

over freedom and autonomy in AI systems. As AI technologies become integral to various aspects of daily life, concerns arise regarding the extent to which individuals may be influenced or even constrained by automated systems. Issues such as algorithmic decision-making, predictive analytics, and autonomous decision systems raise questions about the preservation of human freedom and autonomy. There is a delicate balance to be struck between leveraging AI for efficiency and innovation and ensuring that individuals retain the ability to make informed choices and maintain control over decisions that affect their lives. Ethical considerations in this realm emphasize the need for transparency, user empowerment, and safeguards to prevent undue influence or manipulation, ultimately safeguarding individuals' freedom and autonomy in the age of AI.

Control: The concept of control extends to the optimal verdicts made by AI systems (Habli et al., 2020). AI must be developed in a way that is compatible with human intelligence and values in order to protect decision-making authority (Habli et al., 2020). Control systems within the arena of AI have significantly impacted the accountability of patients who are prone to potentially harmful. Notably, clinicians face serious challenges in exercising control over AI-generated decisions compounded by a limited understanding of these systems. Informed consent is a concern that is raised by the weak control over AI system recommendations, which also jeopardizes patient confidence in clinical care.

Informed Consent: In the context of AI systems, informed consent traditionally pertains to a broad agreement for treatment, specific procedures, or research participation. The examination of nine selected sources sheds light on how informed consent is intricately linked to the management of data within AI frameworks (Wangmo et al., 2019). Firstly, within the domain of health research, consent becomes a means for individuals to authorize or decline the sharing of their data while comprehending the specifics of data recipients and the sharing protocol. Secondly, it is emphasized that securing consent from subjects for sharing vital information with a third party is obligatory (Wangmo et al., 2019). Thirdly, the opacity inherent in machine learning algorithm systems necessitates consent acquisition due to the storage of sensitive information within the system. Furthermore, the anticipation of diverse AI applications requires a complicated model by considering blanket and tiered consent, based on varying levels of engagement. Mechanisms must also be devised to accommodate individuals declining participation in such endeavors. Lastly, the sources underscore the critical and mandatory nature of obtaining consent for both competent and incompetent users, ensuring respect for their autonomy and privacy when subjected to monitoring technologies. The percentage of the published articles from the different sources for ethical issue two is presented in Fig. 2.9.

Ethical issue three

Data Privacy: Data privacy entails the management of an individual's control over their data related to health information (Luxton, 2014). For

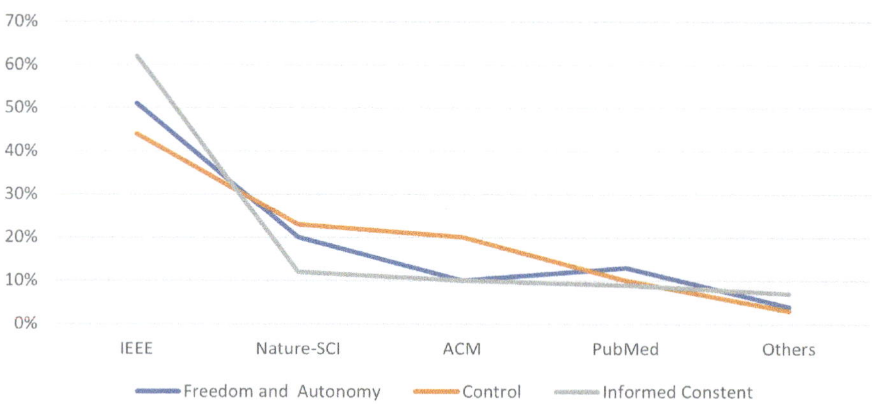

Figure 2.9 Percentage of articles published in different journals for ethical issue two.

instance, a care robot dedicated to vulnerable patients, constantly collects data that is subsequently transmitted to a hospital for medical purposes. In such scenarios, patient privacy rights and self-control may be compromised, including the ability to deny treatment, which is executed by care robots. Issues surrounding data privacy stem from three key factors: data usage, data ownership, and custodianship. The use of private or sensitive health information without the consent of the patient and the possibility of its exploitation for financial gain are two major issues with data usage. Privacy's impact extends beyond mere data protection, influencing users' trust and autonomy when engaging with AI-powered solutions in healthcare. Moreover, various sources highlight the far-reaching consequences of privacy on aspects such as data ownership, discrimination, and overall well-being.

Confidentiality: Confidentiality pertains to the obligation of safeguarding the data privacy of each individual with a high trust factor (Khan et al., 2022). Analogous to the privacy of the data, confidentiality encompasses considerations of the extent to its suitable storage of limited access of sensitive data (Nebeker et al., 2019). The sources that were evaluated indicate that the biggest effect of using AI is the security risk that affects confidentiality. Notably, patient data becomes vulnerable to security breaches or privacy assaults when it is shared or sent to third parties. In research on mobile disorder detection systems, for example, signals are captured using mobile devices and sent to external servers for analysis. The results are then shown and shared with users (Nebeker et al., 2019). The user data that has been gathered is then kept in a database, making it vulnerable to hacking and unauthorized access concerns. The percentage of the articles published in different journals is depicted in Fig. 2.10.

Ethical issue four

Transparency: The ability to understand how an AI system makes decisions is referred to as transparency (Sen et al., 2020). The intrinsic opaqueness of the majority of AI algorithms poses difficulties in comprehending the underlying workings of particular AI methods, especially those related to deep learning. Most remarkably, the fundamental procedures of analysis in the black box and the process of making decisions are still unknown. These difficulties are made worse when algorithms are trained on skewed data or don't include specific demographic traits. For example, a biased risk score was found in an AI system used in the US to forecast the likelihood of recidivism for those who have been accused; an African

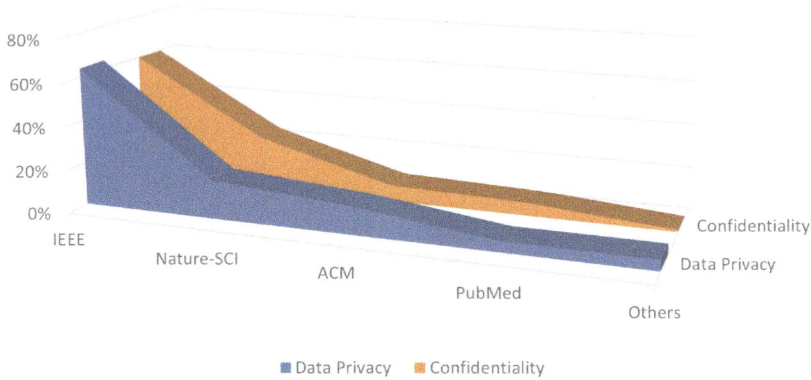

Figure 2.10 Percentage of the articles published in the different journals for ethical issue three.

American accused of little crimes, for example, scored higher than a white American accused of several felonies. Transparency is essential for the development of trust, both among patients and the general public, as well as for the effective application of AI in healthcare. Furthermore, there are further transparency-related considerations when handling sensitive data, including specific medical information.

Explainability: Comparable with transparency, explainability struggles with the opaque character of machine learning (ML) and artificial intelligence (AI) algorithms, leading to difficulties in explaining and understanding the complex link between input data and outputs. User trust is directly impacted by the explainability of AI systems, especially in the medical field where a more thorough and understandable explanation encourages greater confidence in the usage of AI (Vellido, 2019). In the absence of explainability, medical professionals face hurdles in ensuring the credibility of a system and instilling trust in a decision that remains inexplicable, be it to a patient or another healthcare professional (Vellido, 2019). The percentage of the published journals concerning Ethical Issue 4 has been depicted in Fig. 2.11.

Ethical issue five

CyberSecurity: The cybersecurity focus predominantly revolves around the judiciousness, safety measures, risk assessment, and technical resilience of the environment (Verde et al., 2019). Moreover, it is intricately linked to the capacity to take precautions, thereby averting undesirable outcomes and mitigating existential risks (Verde et al., 2019). Notable findings from

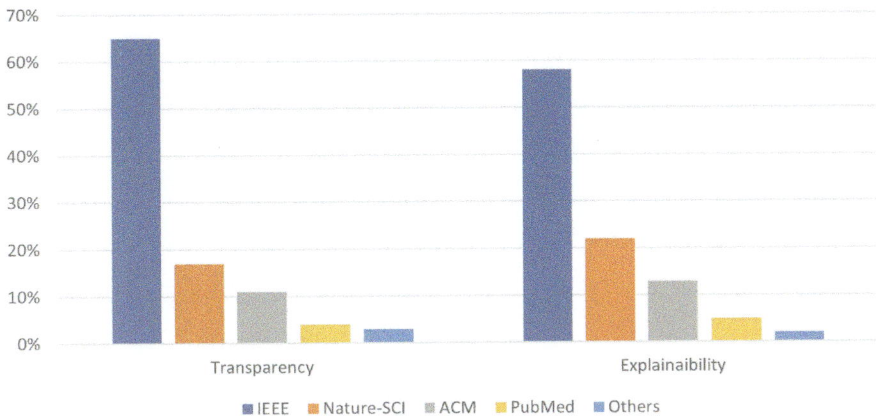

Figure 2.11 Percentage of the articles published in different journals for ethical issue four.

the chosen documents emphasize how important it is for the mental health care sector to give careful thought to issues such as data security of devices that interact with personal health information, data generation methodologies, and potential openings to hacking and unauthorized surveillance. The utilization of advanced tools and techniques for extracting diverse and extensive data from citizens raises the specter of security attacks and privacy invasions. However, these same advanced tools and techniques play a pivotal role in supporting various aspects of data management, including collection, storage, and transmission. They contribute valuable insights for intelligent planning, model development, and innovative data management methods. The percentage of the published articles in the different journals concerning ethical issue 5 has been shown in Fig. 2.12.

Ethical issue six

Trust: Trust embodies a relational and normative concept infused with an element of uncertainty and risk, particularly in delegating tasks to human agents with the inherent concern that they may not unfold as anticipated (Séroussi et al., 2020). The concept of trust plays a crucial role in the therapeutic relationship between patients and human caregivers, and it is as important in interactions between patients and artificial intelligence in healthcare (Séroussi et al., 2020). Unraveling the dynamics of trust in this interaction becomes paramount, with the level of trust contingent upon various facets of data, such as usage in data with technologically driven, which act as the potential for breaches. The dual-use nature of technology

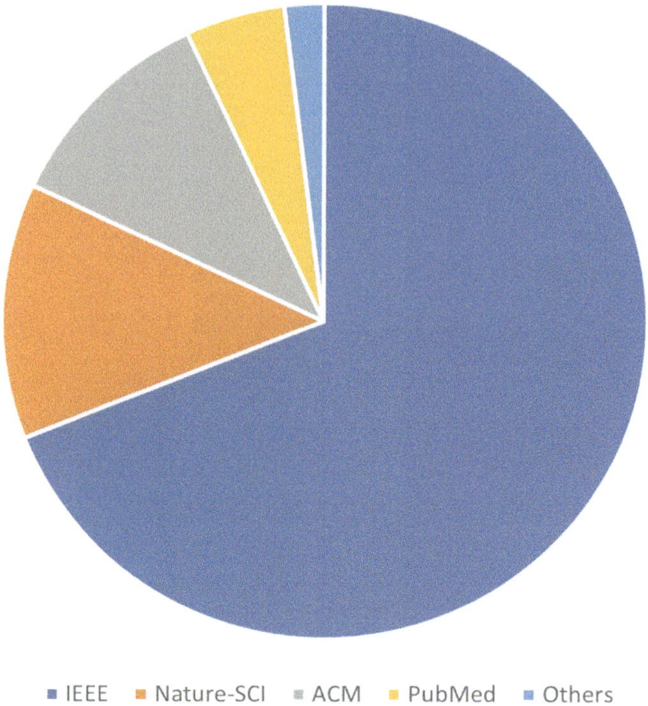

Figure 2.12 Percentage of published articles for ethical issue five.

introduces a layer of complexity, posing a threat to trust in the system and the associated professions, given the multifaceted applications of the technology and the risk that collected data may be repurposed. Trust in AI systems employed in healthcare is also intricately linked to the presence of bias, as AI may produce biased and overfitted outcomes that elude detection by clinicians, consequently eroding user trust and acceptance. Similarly, the issuance of automatic recommendations that lack precision, explainability, and transparency poses a direct threat to patient trust.

Beneficence: Beneficence involves a dedication to acting in the best interest of others, driven by a genuine commitment to improve their well-being (Jacobson et al., 2020). In healthcare, this principle materializes through the actions of healthcare professionals who actively contribute to the welfare and improvement of those under their care (Jacobson et al., 2020). This encompasses efforts to enhance the health and well-being of individuals, minimizing risks, and preventing health issues and illnesses. Additionally, beneficence incorporates ethical considerations, including the careful evaluation of the benefits of interventions in comparison to associated risks and costs. The ethical discourse surrounding beneficence also engages with the moral assessment of deception. While lying is usually considered unethical, there are instances in which it is morally acceptable to advance a patient's physical and emotional well-being. Notably, insights from interviews with individuals with dementia underscore a nuanced acceptance of lying when it aligns with their best interests.

Responsibility: Responsibility in the context of AI systems applied in healthcare pertains to being accountable for the decisions these systems make [47It raises the important topic of determining who is accountable for mistakes or improper performance displayed by AI systems (Raymond Geis et al., 2019). The discourse often revolves around the notion of responsible AI or ML. Beyond considerations of robustness and interpretability, responsible ML emerges as a pivotal factor influencing the adoption of machine learning in healthcare. Accountability for patient harm becomes a necessary facet of AI systems in healthcare. However, the allocation of responsibility raises complexities when AI systems play a role in decision-making. The extent to which human clinicians are held accountable for the harm caused by AI decisions remains unclear, particularly when clinicians may lack direct control over these decisions or when AI systems operate nontransparently. This opacity may make it challenging for clinicians to comprehend how the algorithm determines output based on input data. Furthermore, the creator of AI bears the accountability for guaranteeing that the AI systems' decision-making adheres to the concept of "not harm," and for any harm that may arise from these judgments. Percentage of the articles published in different journals has been shown in.

Supplementary ethical issue

Solidarity: When healthcare adopts AI-powered solutions, solidarity becomes a prominent theme in discussions about justice and equality. Within the domain of justice and equality, prevalent allocation algorithms in healthcare may unintentionally give rise to discrimination against specific groups and races, fostering disparities in resource allocation concerning geography and social economy. Notably, insurers might modify class structures based on risk levels, resulting in stratification among customers. This stratification could translate into unequal treatment for African American patients and disabled individuals compared with their nondisabled Caucasian counterparts. Furthermore, doctors may use factors other than medical urgency to determine which patients should come first.

Concerns about public health are influenced by solidarity, which is closely related to the unequal allocation of care in society and the difficulties carried by caregivers. Algorithms' predictive quality might unintentionally result in algorithmic profiling, which classifies new categories within populations without the subjects' knowledge and endangers the autonomy and uniqueness of patients in social settings. In this setting, highlighting solidarity becomes a useful strategy for reinforcing the community-based aspect of healthcare and highlighting the significance of seeking the common good. Relationships between patients and doctors are essential to the problem of solidarity and are critical to improving treatment results.

Sustainability: The extant research delves deeply into five primary concerns about the sustainability of AI solutions in the healthcare industry. These concerns are categorized as follows: competing interests, divergent contexts, risks and uncertainties, opportunity costs, and democratic inadequacies. In particular, the investigation of sustainability in employability goes into the application of computer algorithms and machine learning (ML) to support occupational health decision-making. The potential for ML to sustain employability is rooted in the development of sophisticated decision support tools, particularly through harnessing data within ML algorithms to forecast suitable interventions. However, the deployment of ML decision-support tools in occupational health raises concerns regarding potential group profiling and discrimination, with the potential to exacerbate social and economic inequalities. The overarching goal is for AI to actively contribute to the sustainable advancement of society, fostering individual health and well-being. Furthermore, the principle of beneficence plays a pivotal role in maintaining sustainability by prioritizing well-being and upholding dignity. Considerations related to health and well-being, coupled with efforts to address disparities between urban and rural healthcare services, enrich the broader conversation on sustainability. Although digital technologies play a crucial role in promoting universal health coverage and advancing Sustainable Development Goals (SDGs), the digital divide presents challenges to sustainability, particularly in low- and middle-income countries where communication barriers persist. The construction of a strong data ecosystem in the field of digital health and the financial viability of the healthcare system are also emphasized as essential elements guaranteeing sustainability. The percentage of the articles related to the supplementary ethical issues is depicted in Fig. 2.13.

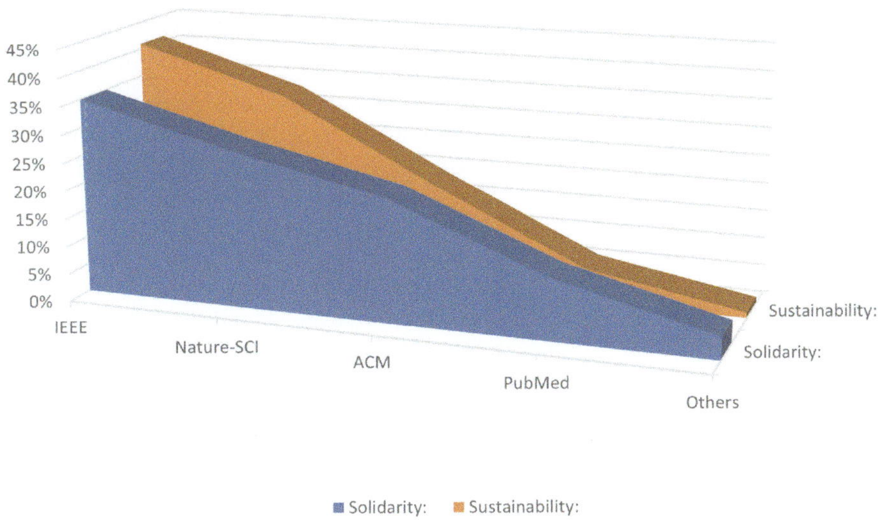

Figure 2.13 Percentage of the articles published in different journals for supplementary issue.

Strategies for solving ethical issues

In this section, various researchers have individually contributed distinct ethical issues, as elucidated above. These contributions are organized into Table 2.2 to clarify the criteria used for their selection. Table 2.2 deals with some of the important contributions by the various researchers, such as integrating different AI applications in nuclear medicine, health equity index based on the applications of ML, and trade-offs at a level in decision-making in different locations.

For ethical issue 2, the contributions of various researchers have emphasized the different biases especially in the radiology data, security data as well as the selection of different datasets, which are used for procrastination purposes are mentioned in Table 2.3.

Ethical issue three features the contributions of different researchers, such as interest in protecting the relevant data and also developing some special ethical codes for safeguarding as outlined in Table 2.4.

Ethical issue four outlines the contributions of various researchers, such as informed consent for digital phenotyping, ethics for heath informatics, and the application of blockchain technology in precision in medicinal data as described in detail in Table 2.5.

Table 2.2 Contributions of the researchers to ethical issue one.

Author	Ethical issue	Contribution
Vellido (2019)	1	Discussed seven issues such as fairness, privacy, animosity, explainability, interpretability, ethics, litigability, and the impact of digital technologies.
Currie et al. (2020)	1	Integration of AI solutions in nuclear medicine for deeper interaction between patient and physician through virtual and physical assessments. The data has been secured through blockchain technology.
Rajkomar et al. (2018)	1	Heath equity is valued, which is based on the fairness index by applying machine learning algorithms. A participatory process has been set up to have equal distribution in the fairness index.
Grote and Berens (2020)	1	Decision-making capabilities for trade-off at the epistemic level, which is at privacy at various locations.

Table 2.3 Contributions of the researchers for ethical issue two.

Author	Ethical issue	Contribution
Caplan et al. (1999)	2	The model has been designed by taking into consideration different biases in the data, and interaction with the predictions with different models.
Raymond Geis et al. (2019)	2	This paper describes radiology in assessing and acquiring the data, which includes informed consent, privacy, and protection of data.
Yeng et al. (2019)	2	The most appropriate observational measure for security purposes and modeling of empirical data obtained from healthcare staff.
Reddy et al. (2020)	2	Governance model where the ethical and regulatory issues are addressed in the healthcare industry.
Beil et al. (2019)	2	Procrastination models are proposed for the selection of datasets and algorithms giving justice to induvial or groups.

Table 2.4 Contributions of the researchers to ethical issue three.

Author	Ethical issue	Contribution
Balthazar et al. (2018)	3	Interest in the ethical manners for protecting the data redundancy of the patients.
Yew (2021)	3	Corebots are designed to take care account of ethics by sufficiently giving the principles of concrete and closely associated with the trust maintenance
Luxton (2014)	3	Ethical issues are described by the care providers with the special recommendation in developing special ethical codes.

Table 2.5 Contributions of the researchers to ethical issue four.

Author	Ethical issue	Contribution
Martinez-Martin et al. (2018)	4	Digital phenotyping is used for informed consent in regarding research studies, which involves a huge amount of data and categorization of heath assessment data.
Séroussi et al. (2020)	4	Ethics are defined for health informatics, which comply with biomedical informatics. AI decision tools are formulated for privacy in data privacy.
Velmovitsky et al. (2021)	4	Precision medicine data is ethically stored by using blockchain technology with a tamperproof concerning the industry. A digital print is stored in the healthcare database for giving security to data sharing.

In Table 2.6, the contributions of different researchers, such as applying different ML algorithms in developing new solutions in the health care industry and also AI based methods are developed especially in the radiology data, which are mentioned in relation to ethical issue 2.

Table 2.7 is detailed in outlining the respective contributions of diverse researchers.

Applications of AI in the health domain

The utilization of electronic health records (EHR) data in AI-driven health research is pivotal. However, the effectiveness of such data hinges on the

Table 2.6 Contributions of the researchers to ethical issue five.

Author	Ethical issue	Contribution
Wiens et al. (2019)	5	The development of ML solutions helps in protecting the challenges in the data and its biases. Specific recommendations are made for deploying new solutions in ethics in the health care industry.
Raymond Geis et al. (2019)	5	Assessing data in radiology when the data is presented for commercial purposes. AI-based techniques are developed to ethically capitalize on the data that can harm the patients.
Buruk et al. (2020)	5	Three documents related to ethics guidelines and the development of AI have been analyzed for the ethical principles and the values they developed. Ethical reasoning has been an important criteria for addressing conflicts.

Table 2.7 Contributions of the researchers to ethical issue six.

Author	Ethical issue	Contribution
Reddy et al. (2020)	6	Ethical and regulatory aspects in the health care industry in the near future. Different types of biases and liability issues had been discussed with its data privacy.
Habli et al. (2020)	6	Moral accountability is where the optimal decisions have been considered for ethical purposes. AI-based systems had been developed for traditional conditions in the robust model.
D'Antonoli et al. (2019)	6	Potential roles of AI have been encountered by considering different biases that are trained examples. A higher score of risk is formulated, which is encountered by analyzing 7000 algorithms generated score.

integrity of the underlying information technology system and database, as heterogeneous or low-quality data can pose challenges (Singh and Kant Singh, 2024). Nevertheless, when AI is appropriately developed and trained with sufficient data, it proves invaluable for scientific exploration, enhancing quality of care, and optimizing clinical practices.

Rather than following the conventional trajectory of scientific publishing, guideline creation, and clinical support tools, well-crafted and well-trained AI can unearth clinical best practices from EHR. Through the analysis of trends in clinical practices gleaned from electronic health data, AI also plays a pivotal role in shaping new models for healthcare delivery (Singh et al., 2023a,b). This extends beyond the routine applications of AI in healthcare, opening avenues for deeper insights into clinical methodologies, and ultimately contributing to the evolution and improvement of healthcare practices.

In the coming years, AI is anticipated to streamline and expedite the pharmaceutical development process. By employing robotics and sophisticated models encompassing genetic targets, drugs, organs, diseases, pharmacokinetics, safety, and efficacy, AI has the potential to transform drug discovery from a labor-intensive to a capital- and data-intensive endeavor. This technology proves instrumental in accelerating the drug development process, making it not only more cost-effective but also highly efficient (Singh et al., 2021; Goyal et al., 2023). The application of AI in drug discovery involves swift identification and assessment of potential drug candidates, significantly reducing the time and resources traditionally required. It doesn't guarantee the automatic success of a therapy, as challenges persist in the development phase. However, AI has demonstrated its capability in identifying potential medicines for challenging conditions such as the Ebola virus. This highlights the promising role of AI in navigating the complexities of drug development, contributing to the advancement of innovative and effective pharmaceutical solutions (Singh et al., 2023).

The ongoing discourse revolves around whether artificial intelligence (AI) fits into existing legal frameworks or warrants the establishment of a distinct category. While AI presents significant opportunities for advancement in healthcare, it brings forth ethical considerations that demand proactive resolutions. The full realization of AI's potential in healthcare hinges on addressing four critical ethical issues (Panesar and Panesar, 2021): securing informed consent for data usage (Das et al., 2021), ensuring safety and transparency (Das et al., 2020), rectifying algorithmic biases and promoting fairness, and (Beck et al., 1996) safeguarding data privacy. The complexity extends beyond legal discussions, delving into politically contentious territory, as highlighted by a resolution of the European Parliament in 2017. Policymakers are striving to proactively tackle the moral challenges associated with the implementation of AI in healthcare. Transparency limitations in AI algorithms raise concerns, particularly in high-risk scenarios, necessitating the development of accountable, equitable, and transparent AI design and governance. However, the very nature of AI processes, often characterized by opacity, complicates the ability to scrutinize them, potentially impeding accountability in cases of harm. This technological gap poses a threat to societal moral frameworks and challenges established legal liability principles.

Artificial intelligence systems (AIS), especially those used in clinical decision-making such as IBM's Watson for oncology, introduce an additional layer of complexity. These systems can make the reasoning behind their outputs opaque, making it challenging to comprehend their decision-making processes. The advent of Machine Learning Healthcare Applications (ML-HCAs) further diversifies functionalities, ranging from self-sufficient diagnostics to predictions about resource allocation. Ensuring the safety and ethical use of these applications in the face of continuous disruptions is imperative.

AI in healthcare requires ongoing adaptation to a dynamic environment while upholding ethical principles. A significant challenge lies in comprehensively understanding AI software failures, particularly when they operate as "black boxes." Researchers must articulate how AI outputs and predictions factor into clinical trials, contributing to cost assessment and guiding further research. Addressing these ethical and practical dimensions is crucial for the responsible and effective integration of AI in healthcare, balancing innovation with the preservation of ethical values and patient well-being.

Conclusion

In undertaking this systematic literature review, we aimed to offer a comprehensive overview of the ethical challenges and strategies relevant to the implementation of AI solutions in healthcare. The identified strategies serve as guidance for developers, policymakers, healthcare institutions, and various stakeholders in the healthcare ecosystem, providing proactive measures for addressing and managing ethical concerns throughout the design processes associated with AI integration. Our initial framework for categorizing ethical issues was drawn from the considerations outlined in two documents, namely, the guidelines by Jobin's group and EGTAI. Through a thematic code mapping process grounded in an abductive approach that incorporates both inductive and deductive methods, we refined the list of ethical issues related to AI. This refinement involved the addition and renaming of issues, informed by insights from Jobin's work and cross-referencing with EGTAI. In our systematic literature review spanning the past decade, we intentionally excluded AI solutions in healthcare developed within the last 3 years due to the ongoing COVID-19 pandemic. However, the global impact of COVID and the varying degrees of lockdown measures have accelerated the adoption of digital healthcare tools, shifting their status from promising opportunities to necessities in many instances. Notably, advancements such as humanoid robots equipped with cognitive technologies, designed to emulate artificial minds for the care of vulnerable patients, have emerged. Given this rapidly evolving landscape, there is a growing urgency to delve into the ethical considerations associated with AI in healthcare and explore corresponding strategic approaches.

Jobin's research and the EGTAI significantly contribute to a thorough understanding of AI applications and their ethical implications across various domains. However, our primary focus in this literature review revolves around the ethical considerations associated with the integration of AI in healthcare. It's essential to acknowledge that the findings we present may not have universal applicability to other domains. This limitation arises from the nuanced definitions of ethical issues when conducting similar reviews in diverse fields or databases. As a result, future efforts should delve into the ethical dimensions of AI utilization in other domains. Additionally, the insights and strategies concerning ethical issues compiled in this paper have the potential for expansion to provide valuable guidance for designers and stakeholders involved in the development of AI solutions, specifically within the healthcare domain.

References

Ahmad, M.A., Patel, A., Eckert, C., Kumar, V., Teredesai, A., 2020. Fairness in machine learning for healthcare. Proc. ACM SIGKDD Int. Conf. Knowled. Discov. Data Min. 3529–3530. https://doi.org/10.1145/3394486.3406461.

Amann, J., Blasimme, A., Vayena, E., Frey, D., Madai, V.I., 2020. Explainability for artificial intelligence in healthcare: a multidisciplinary perspective. BMC Med. Inf. Decis. Making 20, 1–9.

Artificial intelligence and medical imaging 2018: French radiology community white paper. Diag. Interven. Imag. 99 (11), 2018, 727–742. https://doi.org/10.1016/j.diii.2018.10.003.

Badillo, S., Banfai, B., Birzele, F., Davydov, I.I., Hutchinson, L., Kam-Thong, T., Siebourg-Polster, J., Steiert, B., Zhang, J.D., 2020. An introduction to machine learning. Clin. Pharmacol. Ther. 107 (4), 871–885. https://doi.org/10.1002/cpt.1796.

Bærøe, K., Miyata-Sturm, A., Henden, E., 2020. How to achieve trustworthy artificial intelligence for health. Bull. World Health Organ. 98 (4), 257–262. https://doi.org/10.2471/blt.19.237289.

Balthazar, P., Harri, P., Prater, A., Safdar, N.M., 2018. Protecting your patients' interests in the era of big data, artificial intelligence, and predictive analytics. J. Am. Coll. Radiol. 15 (3), 580–586. https://doi.org/10.1016/j.jacr.2017.11.035.

Beck, J., Stern, M., Haugsjaa, E., 1996. Applications of AI in education. XRDS: Crossroads ACM Mag. Stud. 3 (1), 11–15. https://doi.org/10.1145/332148.332153.

Beil, M., Proft, I., van Heerden, D., Sviri, S., van Heerden, P.V., 2019. Ethical considerations about artificial intelligence for prognostication in intensive care. Int. Care Med. Exp. 7 (1). https://doi.org/10.1186/s40635-019-0286-6.

Buruk, B., Ekmekci, P.E., Arda, B., 2020. A critical perspective on guidelines for responsible and trustworthy artificial intelligence. Med. Healthc. Philos. 23 (3), 387–399. https://doi.org/10.1007/s11019-020-09948-1.

Caplan, R.L., Light, D.W., Daniels, N., 1999. Benchmarks of fairness: a moral framework for assessing equity. Int. J. Health Serv. 29 (4), 853–869. https://doi.org/10.2190/DBBU-WUC4-Y23L-4LEA.

Carabantes, M., 2020. Black-box artificial intelligence: an epistemological and critical analysis. AI Soc. 35 (2), 309–317. https://doi.org/10.1007/s00146-019-00888-w.

Currie, G., Hawk, K.E., Rohren, E.M., 2020. Ethical principles for the application of artificial intelligence (AI) in nuclear medicine. Eur. J. Nucl. Med. Mol. Imag. 47 (4), 748–752. https://doi.org/10.1007/s00259-020-04678-1.

D'Antonoli, T.A., Weikert, T.J., Sauter, A.W., Sommer, G., Stieltjes, B., 2019, January. Ethical considerations for artificial intelligence implementation in radiology. European Congress of Radiology-ECR 2019.

Das, S., Arvind, P., Chakraborty, S., Kumari, R., Kumar, S., 2020. IoT-based solar smart tackle free AGVs for industry 4.0. Int. Conf. Int. Things Connec. Technol. 1–7.

Das, S., Kumari, R., Deepak Kumar, S., 2021. A Review on Applications of Simultaneous Localization and Mapping Method in Autonomous Vehicles. Springer Science and Business Media LLC, pp. 367–375. https://doi.org/10.1007/978-981-15-9956-9_37.

Dixit, A., Quaglietta, J., Gaulton, C., 2021. Preparing for the future: how organizations can prepare boards, leaders, and risk managers for artificial intelligence. Healthc. Manag. Forum 34 (6), 346–352. https://doi.org/10.1177/08404704211037995.

Esteva, A., Kuprel, B., Novoa, R.A., Ko, J., Swetter, S.M., Blau, H.M., Thrun, S., 2017. Dermatologist-level classification of skin cancer with deep neural networks. Nature 542 (7639), 115–118. https://doi.org/10.1038/nature21056.

Gerke, S., Minssen, T., Cohen, G., 2020. Ethical and legal challenges of artificial intelligence-driven healthcare. Artif. Intell. Healthc. 295–336. https://doi.org/10.1016/B978-0-12-818438-7.00012-5.

Gibney, E., 2020. The battle for ethical AI at the World's biggest machine-learning conference. Nature 577 (7792). https://doi.org/10.1038/d41586-020-00160-y.

Goyal, L., Dhull, A., Singh, A., Kukreja, S., Singh, K.K., 2023. VGG-COVIDNet: a novel model for COVID detection from X-ray and CT scan images. Proc. Comput. Sci. 218, 1926–1935. https://doi.org/10.1016/j.procs.2023.01.169.

Grote, T., Berens, P., 2020. On the ethics of algorithmic decision-making in healthcare. J. Med. Ethics 46 (3), 205–211. https://doi.org/10.1136/medethics-2019-105586.

Habli, I., Lawton, T., Porter, Z., 2020. Artificial intelligence in health care: accountability and safety. Bull. World Health Organ. 98 (4), 251–256. https://doi.org/10.2471/blt.19.237487.

Hofman, J.M., Sharma, A., Watts, D.J., 2017. Prediction and explanation in social systems. Science 355 (6324), 486–488. https://doi.org/10.1126/science.aal3856.

Jacobson, N.C., Bentley, K.H., Walton, A., Wang, S.B., Fortgang, R.G., Millner, A.J., Coombs, G., Rodman, A.M., Coppersmith, D.D.L., 2020. Ethical dilemmas posed by mobile health and machine learning in psychiatry research. Bull. World Health Organ. 98 (4), 270–276. https://doi.org/10.2471/BLT.19.237107.

Jain, A., Brooks, J.R., Alford, C.C., Chang, C.S., Mueller, N.M., Umscheid, C.A., Bierman, A.S., 2023. Awareness of racial and ethnic bias and potential solutions to address bias with use of health care algorithms. JAMA Health Forum 4 (6). https://doi.org/10.1001/jamahealthforum.2023.1197.

Khan, A.A., Badshah, S., Liang, P., Waseem, M., Khan, B., Ahmad, A., Fahmideh, M., Niazi, M., Akbar, M.A., 2022. Ethics of AI: a systematic literature review of principles and challenges. ACM Int. Conf. Proc. Series 383–392. https://doi.org/10.1145/3530019.3531329.

Kumari, R., Goel, S., Das, S., 2022a. A 3D convolutional neural network approach for diagnosing Alzheimer's disease using modified owl search optimization technique*. IEEE Reg. 10 Ann. Int. Conf., Proc./TENCON 2022. https://doi.org/10.1109/TENCON55691.2022.9977604.

Kumari, R., Goel, S., Das, S., 2022b. Using SVM for Alzheimer's disease detection from 3D T1MRI. MELECON 2022 - IEEE Mediter. Electrotech. Conf. Proc. 600–604. https://doi.org/10.1109/MELECON53508.2022.9842935.

Kumari, R., Das, S., Nigam, A., Pushkar, S., 2023a. Patch-based siamese 3D convolutional neural network for early Alzheimer's disease using multi-modal approach. IETE J. Res. 1–19. https://doi.org/10.1080/03772063.2023.2205857.

Kumari, R., Das, S., Singh, R.K., 2023b. Agglomeration of deep learning networks for classifying binary and multiclass classifications using 3D MRI images for early diagnosis of Alzheimer's disease: a feature-node approach. Int. J. Syst. Assur. Eng. Manage. https://doi.org/10.1007/s13198-023-02180-z.

Lee, H., Tajmir, S., Lee, J., Zissen, M., Yeshiwas, B.A., Alkasab, T.K., Choy, G., Do, S., 2017. Fully automated deep learning system for bone age assessment. J. Digit. Imag. 30 (4), 427–441. https://doi.org/10.1007/s10278-017-9955-8.

Li, X., Brødsgaard, K.E., Jacobsen, M., 2010. Redefining Beijing Consensus: ten economic principles. China Econ. J. 2 (3), 297–311. https://doi.org/10.1080/17538960903529535.

Luxton, D.D., 2014. Recommendations for the ethical use and design of artificial intelligent care providers. Artif. Intell. Med. 62 (1), 1–10. https://doi.org/10.1016/j.artmed.2014.06.004.

Madiega, T.A., 2019. EU Guidelines on Ethics in Artificial Intelligence: Context and Implementation.

Martinez-Martin, N., Insel, T.R., Dagum, P., Greely, H.T., Cho, M.K., 2018. Data mining for health: staking out the ethical territory of digital phenotyping. Npj Digital Med. 1 (1). https://doi.org/10.1038/s41746-018-0075-8.

Morley, J., Machado, C.C.V., Burr, C., Cowls, J., Joshi, I., Taddeo, M., Floridi, L., 2020. The ethics of AI in health care: a mapping review. Soc. Sci. Med. 260. https://doi.org/10.1016/j.socscimed.2020.113172.

Nebeker, C., Torous, J., Bartlett Ellis, R.J., 2019. Building the case for actionable ethics in digital health research supported by artificial intelligence. BMC Med. 17 (1). https://doi.org/10.1186/s12916-019-1377-7.

Ouchchy, L., Coin, A., Dubljević, V., 2020. AI in the headlines: the portrayal of the ethical issues of artificial intelligence in the media. AI Soc. 35 (4), 927–936. https://doi.org/10.1007/s00146-020-00965-5.

Panesar, A., Panesar, H., 2021. Artificial intelligence and machine learning in global healthcare. Handbook of Global Health: With 362 Figures and 152 Tables, pp. 1775–1813. https://doi.org/10.1007/978-3-030-45009-0_75.

Rajkomar, A., Hardt, M., Howell, M.D., Corrado, G., Chin, M.H., 2018. Ensuring fairness in machine learning to advance health equity. Ann. Intern. Med. 169 (12), 866–872. https://doi.org/10.7326/M18-1990.

Raymond Geis, J., Brady, A.P., Wu, C.C., Spencer, J., Ranschaert, E., Jaremko, J.L., Langer, S.G., Kitts, A.B., Birch, J., Shields, W.F., van den Hoven van Genderen, R., Kotter, E., Gichoya, J.W., Cook, T.S., Morgan, M.B., Tang, A., Safdar, N.M., Kohli, M., 2019. Ethics of artificial intelligence in radiology: Summary of the joint European and North American multisociety statement. Radiology 293 (2), 436–440. https://doi.org/10.1148/radiol.2019191586.

Reddy, S., Allan, S., Coghlan, S., Cooper, P., 2020. A governance model for the application of AI in health care. J. Am. Med. Inf. Assoc. 27 (3), 491–497. https://doi.org/10.1093/jamia/ocz192.

Sen, D., Chakrabarti, R., Chatterjee, S., Grewal, D.S., Manrai, K., 2020. Artificial intelligence and the radiologist: the future in the armed forces medical services. BMJ Mil. Health 166 (4), 254–256. https://doi.org/10.1136/jramc-2018-001055.

Séroussi, B., Hollis, K.F., Soualmia, L.F., 2020. Transparency of health informatics processes as the condition of healthcare professionals' and patients' trust and adoption: the rise of ethical requirements. Yearb. Med. Inform. 29 (1), 7–10. https://doi.org/10.1055/s-0040-1702029.

Shin, D., 2021. The effects of explainability and causability on perception, trust, and acceptance: implications for explainable AI. Int. J. Hum. Comput. Stud. 146. https://doi.org/10.1016/j.ijhcs.2020.102551.

Singh, A., Kant Singh, K., 2024. Blockchain and deep learning: research challenges, open problems, and future. Blockchain Deep Learn. Smart Healthc 425–431.

Singh, A., Dhull, A., Singh, K.K., Singh, K.K., 2023a. AIML-Based blockchain solutions for IoMT. Blockchain Deep Learn. Smart Healthc 73–93. https://doi.org/10.1002/9781119792406.ch4.

Singh, A., Dhull, A., Singh, K.K., 2023b. Blockchain with federated learning for secure healthcare applications. Blockchain Deep Learn. Smart Healthc 35–44. https://doi.org/10.1002/9781119792406.ch2.

Singh, K.K., Singh, A., Lin, J.W., Elngar, 2021. Deep Learning and IoT in Healthcare Systems: Paradigms and Applications. CRC Press, 2021.

Srivastava, S., Soman, S., Rai, A., Srivastava, P.K., 2017. Deep learning for health informatics: recent trends and future directions. 2017 Int. Conf. Adv. Comput. Commu. Inform. ICACCI 2017 2017, 1665–1670. https://doi.org/10.1109/ICACCI.2017.8126082.

Stahl, B.C., Antoniou, J., Ryan, M., Macnish, K., Jiya, T., 2022. Organisational responses to the ethical issues of artificial intelligence. AI Soc. 37 (1), 23–37. https://doi.org/10.1007/s00146-021-01148-6.

Toulouse, E., Masseguin, C., Lafont, B., McGurk, G., Harbonn, A., A Roberts, J., Granier, S., Dupeyron, A., Bazin, J.E., 2018. French legal approach to clinical research. Anaesthesia Crit. Care Pain Med. 37 (6), 607–614. https://doi.org/10.1016/j.accpm.2018.10.013.

Vellido, A., 2019. Societal issues concerning the application of artificial intelligence in medicine. Kidney Dis. 5 (1), 11–17. https://doi.org/10.1159/000492428.

Velmovitsky, P.E., Bublitz, F.M., Fadrique, L.X., Morita, P.P., 2021. Blockchain applications in health care and public health: increased transparency. JMIR Med. Inform. 9 (6). https://doi.org/10.2196/20713.

Verde, L., De Pietro, G., Alrashoud, M., Ghoneim, A., Al-Mutib, K.N., Sannino, G., 2019. Leveraging artificial intelligence to improve voice disorder identification through the use of a reliable mobile app. IEEE Access 7, 124048–124054. https://doi.org/10.1109/ACCESS.2019.2938265.

Wangmo, T., Lipps, M., Kressig, R.W., Ienca, M., 2019. Ethical concerns with the use of intelligent assistive technology: findings from a qualitative study with professional stakeholders. BMC Med. Ethics 20 (1). https://doi.org/10.1186/s12910-019-0437-z.

Weller, A., 2019. Transparency: Motivations and Challenges, vol 11700. Springer Science and Business Media LLC, pp. 23–40. https://doi.org/10.1007/978-3-030-28954-6_2.

Wiens, J., Saria, S., Sendak, M., Ghassemi, M., Liu, V.X., Doshi-Velez, F., Jung, K., Heller, K., Kale, D., Saeed, M., Ossorio, P.N., Thadaney-Israni, S., Goldenberg, A., 2019. Do no harm: a roadmap for responsible machine learning for health care. Nat. Med. 25 (9), 1337–1340. https://doi.org/10.1038/s41591-019-0548-6.

Xu, W., 2019. Toward human-centered AI: a perspective from human-computer interaction. Interactions 26 (4), 42–46. https://doi.org/10.1145/3328485.

Yeng, P.K., Yang, B., Snekkenes, E.A., 2019. Observational measures for effective profiling of healthcare staffs' security practices. COMPSAC 2, 397–404. https://doi.org/10.1109/COMPSAC.2019.10239.

Yew, G.C.K., 2021. Trust in and ethical design of carebots: the case for ethics of care. Int. J. Soc. Robot. 13 (4), 629–645. https://doi.org/10.1007/s12369-020-00653-w.

Zeng, J., Ustun, B., Rudin, C., 2017. Interpretable classification models for recidivism prediction. J. Roy. Stat. Soc. Stat. Soc. 180 (3), 689–722. https://doi.org/10.1111/rssa.12227.

CHAPTER 3

Explainable AI methods to increase trustworthiness in healthcare

Dmytro Savchuk and Anastasiya Doroshenko
Lviv Polytechnic National University, Institute of Computer Science and Information Technologies, Department of Automated Control Systems, Lviv, Ukraine

Introduction

Artificial intelligence methods have long been widely used to solve various medical problems: diagnosis, processing of medical images, forecasting of development, and spread of diseases. With the improvement of AI methods and the increase of available computing resources, scientists have managed to achieve an impressively high accuracy in solving these problems. However, the methods that allow us to get the best results, such as deep learning networks, neural networks, and SVM, are "black-box" models (Hohma and Lütge, 2023). Accordingly, although we can guarantee the accuracy of the obtained forecast after testing and validation, we are unable to explain why this particular decision was made.

This has become a particularly urgent problem for developers of AI systems now because the governments of the EU, USA, China, UK, and many other countries have started to regulate the use of AI systems by developing relevant regulations. One of the main requirements stipulated by this regulation is the legality, transparency, nondiscrimination, and reliability of artificial intelligence systems. In this regard, one of the requirements of trustworthy AI is the ability to explain the decisions made by the AI system. Accordingly, there is an actual task to develop and use Explainable AI methods (Hulsen, 2023).

To address the risks associated with the development and utilization of AI, the European Parliament and Council have introduced regulations known as the AI Act. These regulations propose a four-tier risk

Responsible and Explainable Artificial Intelligence in Healthcare
ISBN 978-0-443-24788-0
https://doi.org/10.1016/B978-0-443-24788-0.00003-0

categorization system, classifying AI systems into categories of unacceptable, high, limited, and minimal risk. AI systems falling into the unacceptable risk category will face partial or complete prohibition from use in the EU market. High-risk systems will be subject to conformity assessments, leading to additional restrictive measures, primarily impacting AI providers. Systems posing limited risks will be required to meet specific transparency requirements, while minimal-risk systems will not be subject to binding measures but are encouraged to follow voluntary self-imposed codes of conduct similar to high-risk systems (Bharati et al., 2023).

As AI systems heavily rely on data processing, they are subject to the constraints outlined in the General Data Protection Regulation (GDPR), particularly when dealing with personal data or special categories of data. In such cases, additional safety and privacy requirements must be met, and data subjects are granted specific rights to control their data's handling by data operators. Alongside the GDPR, other EU regulations addressing data, both personal and nonpersonal, can affect AI providers, albeit to a limited extent (Public Voice coalition, Universal Guidelines for Artificial Intelligence, 2018). The Privacy and Electronic Communications Regulation 2017/0003 (ePrivacy Regulation) specifies rules for electronic communications data, complementing the GDPR, particularly when this data is considered personal ("How to use AI and personal data appropriately and lawfully," 2023). The proposal for Regulation 2022/0047 on harmonized rules for data access and use (Data Act) sets obligations for sharing data generated by physical products collecting and transmitting data. Regulation 2020/0340 on European data governance (Data Governance Act) addresses data sharing and reuse among EU stakeholders, enhancing data availability and exchange. While these regulations may impact organizations providing AI, their obligations are more closely tied to the entire system and have indirect implications for the AI component.

Compliance with this requirement is particularly important for medical AI systems, as they process not only the personal data of users but also information belonging to special categories of sensitive data defined by the GDPR (health data, biometrics, and genetic data) ("Regulation (EU) 2022/868 of the European Parliament and of the Council of 30 May 2022 on European data governance and amending Regulation (EU) 2018/1724 (Data Governance Act," 2022).

When AI components are integrated into products, they may be subject to product safety and liability requirements, necessitating protective measures from AI component providers. The General Product Safety Directive

(GPSD) and the Product Liability Directive (PLD) provide mechanisms for enforcing consumer claims related to damage, with PLD outlining liability for claims resulting from product defects or unsafe products, and product safety regulations setting safety specifications for products. The GPSR serves as a safety net for products not covered by the AI Act's specific safety rules, such as products with AI components posing minimal risk ("Proposal for a regulation of the European parliament and of the council on harmonised rules on fair access to and use of data (Data Act)," n.d.). Additionally, the EU is considering an AI Liability Directive (AILD) that addresses claims related to AI-based systems, adapting access to information and the burden of proof to AI's unique circumstances.

To create a safer digital environment that protects user fundamental rights and ensures a level playing field for businesses, the EU has expanded existing regulations in the field of commercial practices. This includes the Digital Services Act (DSA) and the Digital Markets Act (DMA). The DSA imposes restrictions on digital service providers, which can affect AI-based applications, particularly in areas such as online advertising targeting and recommender systems. The DMA places additional obligations on "gatekeepers," large providers of core platform services, including those offering AI-based recommender systems. Existing legal frameworks on commercial practices, such as the Unfair Commercial Practices Directive (UCPD), prohibit misleading and aggressive commercial practices, which can impact AI applications, as AI providers must not engage in deceptive practices regarding product characteristics and compliance with codes of conduct.

Fundamental rights are outlined in the Charter of Fundamental Rights of the European Union legal frameworks. While they do not directly impose obligations on AI providers, they are frequently referenced in the context of AI, as AI's introduction is associated with risks and challenges to the protection of fundamental rights. This places an obligation on the state to safeguard its citizens from fundamental rights violations and an indirect obligation on AI providers to adhere to the specified provisions and measures against fundamental rights restrictions ("Proposal for a regulation of the European parliament and of the council on harmonised rules on fair access to and use of data (Data Act)," n.d.).

In conclusion, EU policy views AI governance in a similar light and has already integrated several measures into regulatory and legal guidelines. The applicability of these measures depends on the specific AI application and the target group involved. So, the main advantage of XAI is compliance with laws and regulations that require transparency and explainability in AI systems.

Methodology

Despite the fact that today AI is widely used in various areas of human activity — industry, education, entertainment, and marketing, there are some problems with the effective use of AI in health care. In particular, for some medical problems, it is quite difficult to collect large amounts of data to form a training sample. Accordingly, when the patient sample has limited diversity in terms of race, ethnicity, gender, age, etc., the results of these studies may be difficult to apply to patients with other characteristics. This may cause some bias inherent in the training sample, which is strictly prohibited by such regulations as, for example, the AI Act. To overcome this problem, data sets using larger and more diverse patient cohorts can be created, or bias can be taken into account when designing experiments.

Another important point to pay attention to when solving medical problems is privacy and security issues, which are strictly regulated today (GDPR, HIPAA, PIPL), the possibility of using personal data is limited, and large fines may be imposed for the leakage of such data. Various approaches can be used to reduce these risks: from the radical one of using synthetic data sets generated on the basis of existing AI algorithms (GAN, Generative Adversarial Networks) to organizing the system architecture in such a way as to minimize the transfer of personal data by training AI directly in a data storage location, for example using a centralized data store.

However, the main problem for the wider use of AI in medicine is the limited trust of doctors and patients in AI algorithms (Yang et al., 2022). Due to the fact that today's most popular machine learning algorithms work according to the "black box" principle, when we only see the input data and the obtained result, users may have doubts about the decision made (Barredo Arrieta et al., 2020). This lack of trust in AI algorithms, their accuracy, and reliability is a very serious problem in a field where a decision can mean life or death for a patient (Sharma et al., 2024). As a result of this mistrust, the question of creating Explainable Artificial Intelligence (XAI) becomes particularly relevant. XAI is focused on making algorithms and resulting decisions understandable to humans (Haresamudram et al., 2023).

In an interpretable model, healthcare professionals have the ability to examine the rationale behind an incorrect outcome and potentially identify any manipulations (Sheu and Pardeshi, 2022). Moreover, the functioning of the AI algorithm can shed light on any systematic biases (Tjoa and Guan, 2021). For instance, it can become evident that the algorithm was solely trained on data from individuals of a specific ethnic background (Goyal et al., 2022). This can lead to discrepancies when the AI algorithm proposes courses of action that are not aligned with an individual patient's unique circumstances (Doroshenko, 2022). The algorithm might misinterpret a correlation as causation, resulting in, for instance, an inaccurate treatment recommendation (Jung et al., 2023). In the case of a black-box AI system, such discrepancies may go unnoticed, but in a transparent and explainable AI model, it becomes much easier to detect or at least highlight the potential risk of such a mismatch (Rai, 2020).

Different models of machine learning have different degrees of possible explainability. Thus, in models belonging to transparent, or so-called white models (linear models, rule-based models, and decision trees), no additional effort is needed to review the decision-making process (Loyola-Gonzalez, 2019; Ribeiro et al., 2016; Gerlings et al., 2022). While black box models such as Convolutional Neural Networks (CNN), Recurrent Neural Networks (RNN), Support Vector Machines (SVM), and ensemble models do not have this capability (Singh and Kant Singh, 2024). Fig. 3.1 shows a graph of the dependence of the accuracy of the most popular AI methods depending on the level of their built-in explainability (Yang et al., 2022). Accordingly, it can be concluded that simple models with good explainability usually demonstrate comparatively lower accuracy than, for example, models based on deep learning (Jangra et al., 2023). Accordingly, the task of the developer of AI systems is to achieve a reasonable compromise between accuracy and explainability (Currie et al., 2019).

In (Yang et al., 2022) defined that the ideal solution should combine high explainability and high accuracy. However, existing linear models, rule-based models, and decision trees are more transparent, but with lower performance in general. In contrast, complex models, e.g., deep learning and ensembles (Shafay et al., 2023), manifest higher performance while less explainability can be obtained (Khurana et al., 2024).

Let's consider the possibilities of increasing the transparency and explainability of decision-making in medical appointment systems using the example of the task of diagnosing heart diseases. We use the most common machine learning and AI methods for this classification task, compare their results, and analyze the relationship between classification accuracy and the level of explainability of decision-making for this method (Doroshenko, 2022).

We will also consider the process of preliminary processing and analysis of training data as a means of increasing the reliability of the AI system.

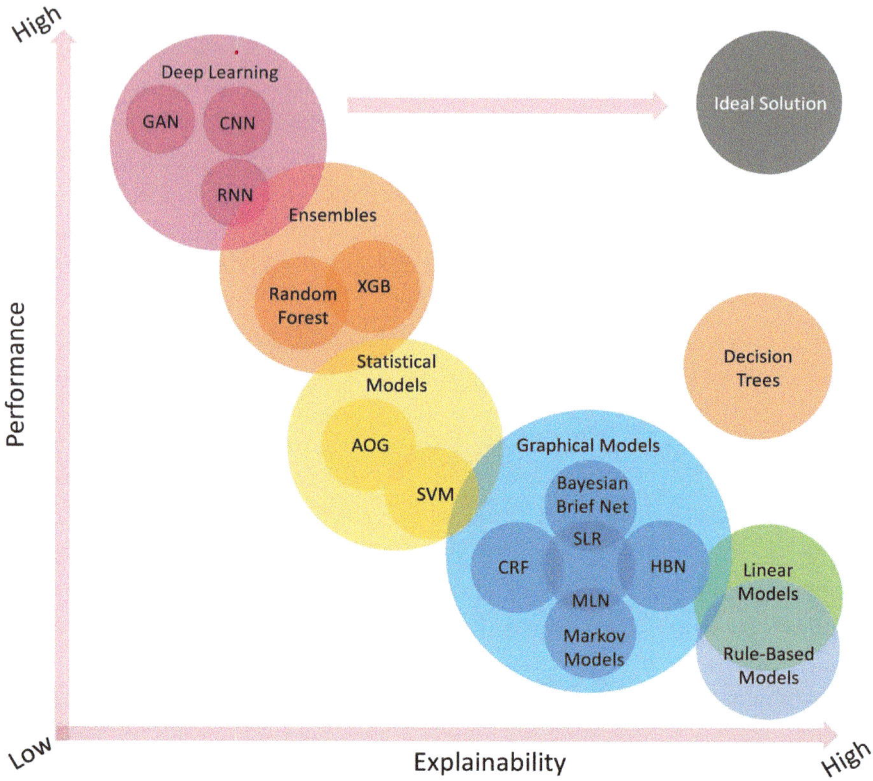

Figure 3.1 Model explainability versus model performance for widely used machine learning and deep learning algorithms.

Data analysis and visualization

Description of the classification task

The rapid increase in the incidence of cardiovascular disease (CVD) is one of the greatest public health challenges of our time. These types of diseases cause the destruction of entire families and reduce the well-being of society. According to the World Health Organization (WHO), cardiovascular diseases claim about 18 million lives annually, accounting for approximately 31% of all deaths worldwide (UCI Machine Learning Repository, 2023).

One of the most dangerous complications of CVD is the possibility of heart failure. This type of condition occurs when the heart cannot pump enough blood as much as the body needs, which usually leads to death. Therefore, there is an urgent need for early detection and treatment of cardiovascular diseases. To address this issue, the best way is to use machine learning models that can improve the effectiveness of treating.

The analysis is based on a dataset that was collected from many individuals (Heart Failure Prediction Dataset, 2021). It was created by combining five independent samples of cardiovascular disease data (Alizadehsani et al., 2019). The total number of people surveyed in the dataset is 1190, which after removing duplicates is 918. The original dataset contained 76 attributes, but after removing redundancies, it is 11.

The main data collection places are as follows.

- Cleveland (303 patients)
- Hungary (294 patients)
- Switzerland (123 patients)
- Long Beach VA (200 patients)
- Stalog (Heart) Data Set (270 patients)

The main questions that were asked to the patient.

Age

Refers to the chronological age of a patient, measured in years. This is an important demographic characteristic in the detection of cardiovascular disease, as age is a recognized risk factor. Older people are usually at a higher risk of cardiovascular disease.

Sex

Refers to the biological or physiological classification of a person as male or female. In cardiovascular medicine, gender, such as age, is considered an important demographic factor. This is due to the fact that the course of cardiovascular disease may differ between men and women.

Chest pain type

It characterizes the discomfort or pain in the chest experienced by the patient. It is often divided into several types, for example, typical angina (substernal chest discomfort with a characteristic quality and duration that is provoked by exertion or emotional stress and relieved by rest or nytroglycerin), atypical angina (cannot be relieved by rest or nitroglycerin), nonanginal pain (meets one or none of the typical angina characteristings), and asymptomatic pain. This test can give you a general idea of the nature of your heart-related symptoms.

Resting BP

It is a blood pressure measured when a patient is resting. It is usually expressed as two readings (Table 3.1) systolic blood pressure (the pressure

Table 3.1 Resting blood pressure.

Category	Systolic BP (mm Hg)		Diastolic BP (mm Hg)
Ideal	<120	AND	<80
Normal	120−129	AND	80−84
Hight normal	130−139	OR	85−89
Hypertension			
Stage 1	140−159	OR	90−99
Stage 2	160−179	OR	100−109
Stage 3	≥180	OR	≥110

when the heart is beating) over diastolic blood pressure (the pressure when the heart is at rest). High blood pressure at rest is a significant risk for heart disease.

Cholesterol

Blood cholesterol level (Table 3.2) − defined as the concentration of cholesterol, which includes low-density lipoprotein (LDL) cholesterol and high-density lipoprotein (HDL) cholesterol. Abnormal cholesterol levels, for example high LDL cholesterol levels, increase the risk of atherosclerosis or heart disease.

Fasting BS

Blood sugar, or in other words blood glucose, is the concentration of glucose in the blood after an overnight fast. Elevated blood sugar levels usually indicate diabetes or impaired glucose metabolism, which in turn leads to a risk of heart disease. Table 3.3 below shows the gradation of blood sugar levels.

Table 3.2 Cholesterol level.

Category (mg/dL)	Total	LDL	HDL	Triglycerides
Ideal	<200	<100	>60	<150
Borderline	200−239	130−159	Women: 40−59 Men: 50−59	150−199
High	>240	High: 160−189 Very high: >190	Women: <40 Men: <50	High: 200−499 Very high: >500

Table 3.3 Blood sugar level.

Category	Fasting	Just ate	Three hours after eating
Normal	80–100	170–200	120–140
Prediabetic	101–125	190–230	140–160
Diabetic	<126	220–300	<200

Resting ECG (electrocardiogram)

The electrical activity of the heart is usually recorded when a person is at rest. It helps to obtain information about heart rate, rhythm, and electrical conduction abnormalities, which can be used to diagnose various heart diseases (Alizadehsani et al., 2019).

The following Resting ECG variants are shown in this dataset (Fig. 3.2): ***normal*** – indicates normal cardiac behavior, ***ST*** – with ST-T wave

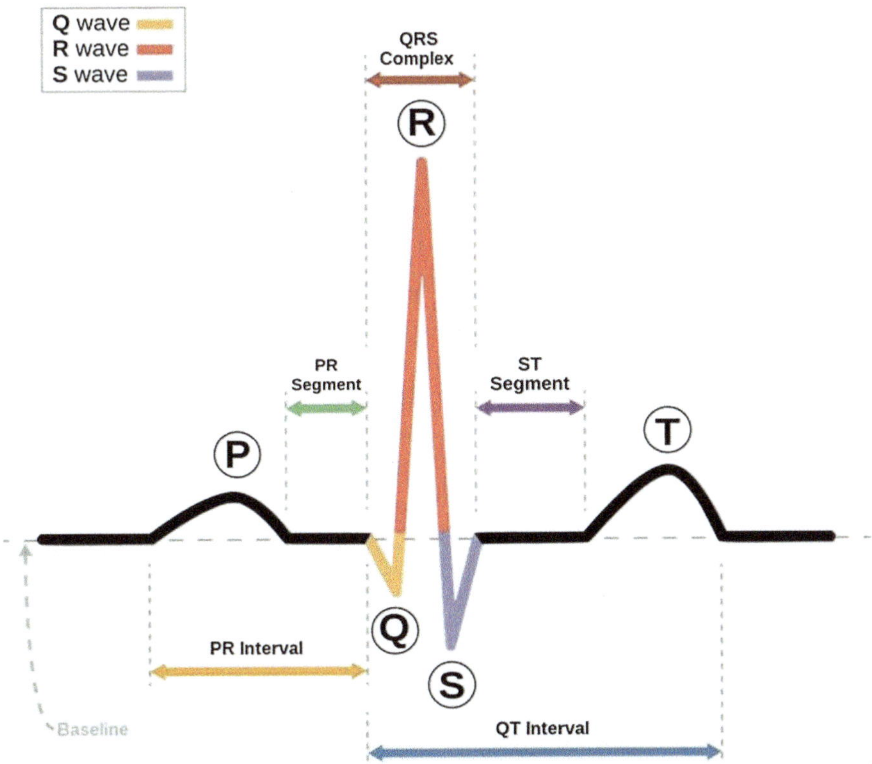

Figure 3.2 Electrocardiogram example.

abnormalities (T wave inversion and/or ST elevation or depression >0.05 mV), ***LVH*** – with probable or definite left ventricular hypertrophy according to Estes' criteria (Alizadehsani et al., 2019).

Max HR

Defined as the highest heart rate that a person can achieve during exercise (Table 3.4). It is usually used to calculate the target heart rate for exercise testing. The maximum heart rate changes with age and is an important parameter in assessing the efficiency of the cardiovascular system.

Exercise angina

Exercise-induced angina, or angina pectoris, is defined as chest pain or discomfort that occurs during physical activity or exertion. It is one of the important symptoms that may indicate coronary heart disease.

Oldpeak

Oldpeak, or in other words ST segment depression, refers to the degree of ST segment depression (in millimeters) on an electrocardiogram during exercise. It is one of the critical parameters used in stress testing to assess the degree of possible myocardial ischemia (reduced blood supply to the heart).

ST slope

It means the slope of the ST segment on an electrocardiogram (ECG) and indicates the direction and degree of deviation from the baseline. It is usually used to assess the presence of myocardial ischemia.

Table 3.4 Maximum heart rate.

Category	% of max HR	Exertion level	Effect
5	90–100	Max	Maximum performance and speed
4	80–90	Hard	Maximum performance capacity
3	70–80	Moderate	Aerobic fitness
2	60–70	Light	Basic endurance and fat burning
1	50–60	Very light	Overall health and helps recovery
0	>50	Rest	No meaningful effect

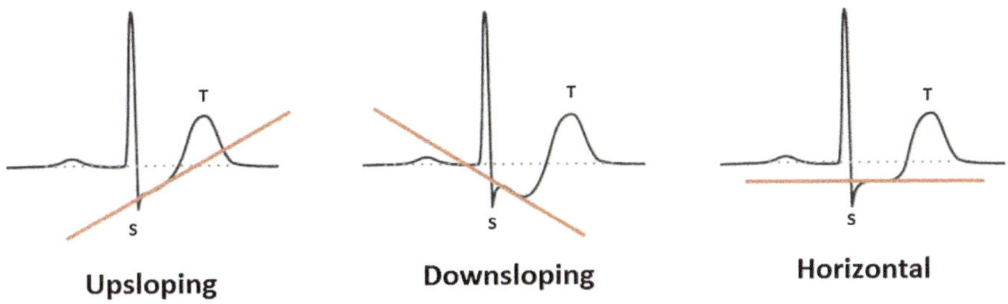

Figure 3.3 ST segment depression.

In Fig. 3.3, we can see the possible variants of Oldpeak present in the dataset. Namely, ***up***: upsloping, ***flat***: horizontal, ***down***: downsloping. These parameters can provide diagnostic information about heart disease.

Data sample

First of all, it is necessary to consider the initial data set. As mentioned above, the dataset was originally composed of five parts and contained 76 characteristics, but after removing outliers, only 11 characteristics remained, as discussed in the previous section (Heart Failure Prediction Dataset, 2021). These included age, gender, type of chest pain, resting blood pressure, cholesterol, fasting blood sugar, resting electrocardiogram, maximum heart rate, angina during exercise, Oldpeak, and ST segment elevation. The total number of observations was 918 without duplicates (Alizadehsani et al., 2019).

Features of the data set

First, let's look at the independent features and their possible variants (Doroshenko, 2019). Specifically, in Table 3.5 below we can see their gradation.

As we can see from the table above, the dataset has both quantitative and qualitative features. Therefore, this should be taken into account when processing the data set and creating models (Savchuk and Doroshenko, 2021).

Table 3.6 shows the possible options of the value of the target class. As we can see, it has only two values: Heart Disease (Hohma and Lütge, 2023) and Normal Condition (0), which corresponds to the binary classification task.

Explaining with data visualization

Now let's look at all the categorical independent features (sex, chest pain type, fasting BS, resting ECG, exercise angina, ST slope) of the dataset. To do this, let's build a histogram where each column corresponds to the number of certain values of the feature in the data set. As a result, we can see the following Fig. 3.4.

Table 3.5 Independent features.

Feature number	Name	Values
X1	Age	Numeric value
X2	Sex	M – man, W – woman
X3	Chest Pain Type	ATA, NAP, ASY, TA
X4	Resting BP	Numeric value
X5	Cholesterol	Numeric value
X6	Fasting BS	0 – No, 1 – Yes
X7	Resting ECG	Normal, ST, LVH
X8	Max HR	Numeric value
X9	Exercise angina	N – No, Y – Yes
X10	Oldpeak	Numeric value
X11	ST Slope	Up, flat, down

Table 3.6 Target class.

Feature number	Name	Values
Y1	Heart Disease	1
Y2	Normal Condition	0

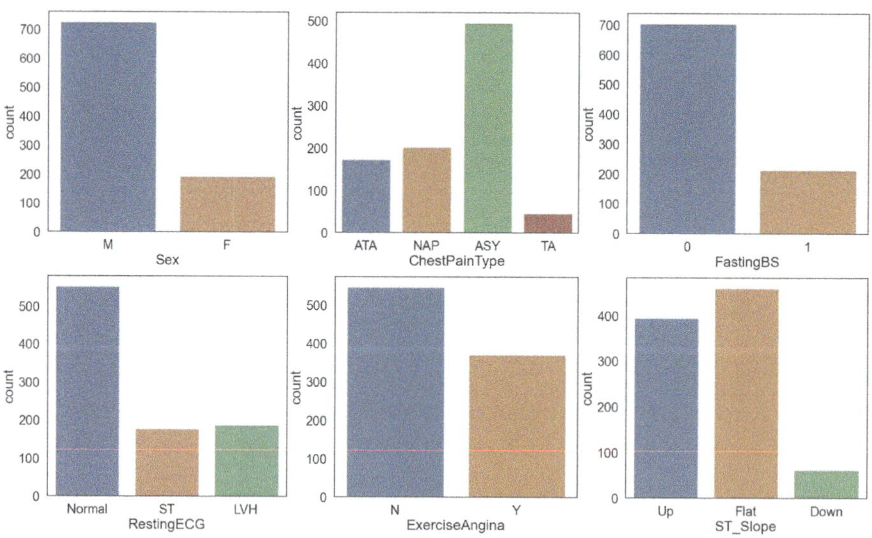

Figure 3.4 Qualitative features.

From the figure above, we can understand the following.

- There are approximately three times as many male patients as female patients.
- The most common type of chest pain is asymptomatic (ASY). Whereas atypical angina (ATA) and nonanginal pain (NAP) account for approximately the same number of patients, and typical angina (TA) is the least.
- There are three times fewer people with high blood sugar than with normal levels.
- There are three times more people with a normal resting electrogram than people with ST-T wave disorders (ST) or left ventricular hypertrophy (LVH).
- The number of patients experiencing and not experiencing angina from exercise is almost the same.
- The number of people with an upward (Up) and flat ST slope is almost the same, while there are almost no people with a downward (Down) slope.

Fig. 3.5 shows a density histogram of patients by age depending on the target class (Heart Desiese and Normal condition). It shows that the average age of patients with heart problems is 57 years, as opposed to 52 years when people usually do not have any diseases. Also, the average age of patients is about 53.5 years with a deviation of 9.5 years, and a skewness of − 0.19 and kurtosis of − 0.39.

In Fig. 3.6, we can see the density distribution based on the patients' blood pressure. In general, the data is evenly distributed for patients with and without heart problems. However, there are a few outliers when the patient's blood pressure is 0. Therefore, these observations should be

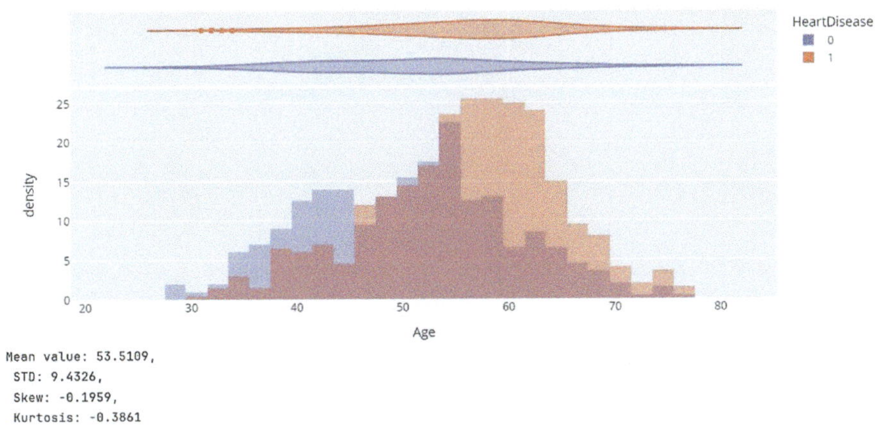

Figure 3.5 Age feature.

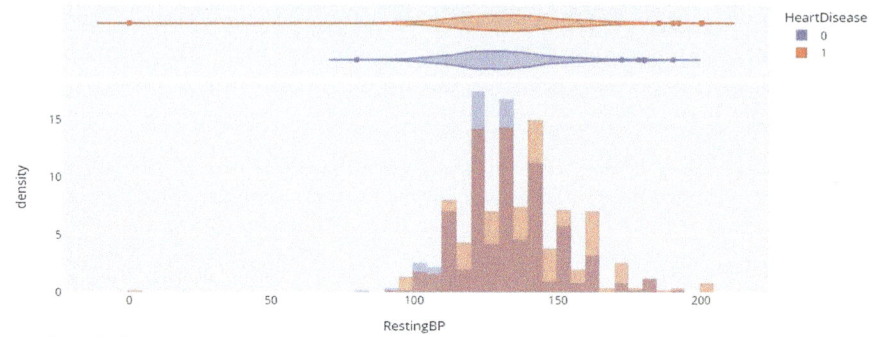


```
Mean value: 132.3965,
 STD: 18.5142,
 Skew: 0.1798,
 Kurtosis: 3.2713
```

Figure 3.6 Resting blood pressure feature.

discarded in the modeling process. The average blood pressure value is 132, which is a normal category, with a standard deviation of 18.5, a skewness coefficient of 0.18, and a Kurtosis of 3.3.

Fig. 3.7 shows that the histogram has quite a few outliers. Namely, when the cholesterol level is 0. They need to be removed in order to make the assessment effective and efficient. Cholesterol levels, such as blood pressure, are fairly evenly distributed for the two types of patients. The mean is 199, which is right on the borderline. The standard deviation is 109, because the number of outliers in the dataset is quite big, so this resulted in a large difference between the threshold values. The skewness is − 0.61 and the kurtosis coefficient is 0.11.

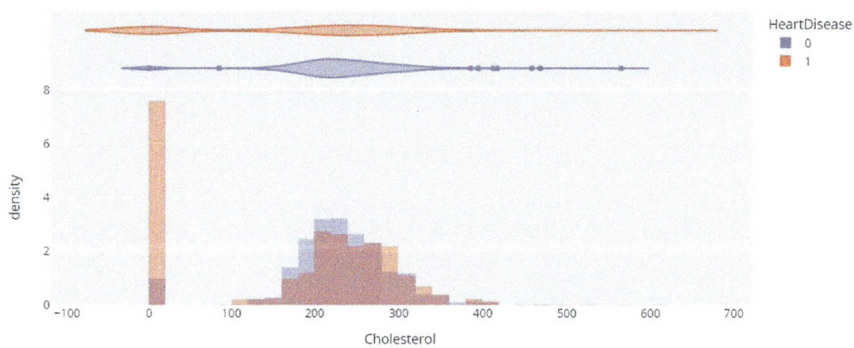


```
Mean value: 198.7996,
 STD: 109.3841,
 Skew: -0.6101,
 Kurtosis: 0.1182
```

Figure 3.7 Cholesterol feature.

Now let's move on to the maximum heart rate. As we can see in Fig. 3.8, the performance of people with and without heart problems is quite different. This can be seen both in the histogram and by the standard deviation of 25.5. Also, the average rate for people with heart problem is 121 heartbeats per minute, while for healthy people it is 153, which in turn gives a mean of 137. Other metrics are − 0.14 (Skewness) and − 0.44 (Kurtosis), respectively.

In the following Fig. 3.9, we can see that the largest number of patients' Oldpeak values are 0, which in turn may mean zero ST-segment depression. The average ST-segment depression value is 0.8 mm with a standard deviation of one and a skewness of one and a Kurtosis coefficient of 1.2.

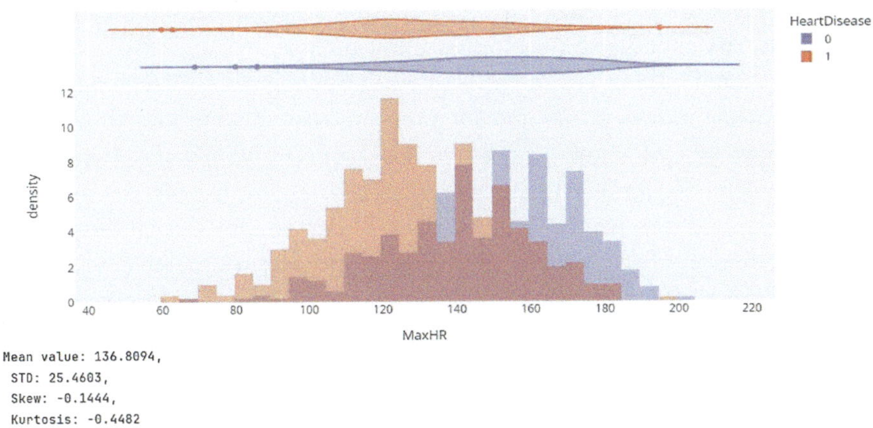

Figure 3.8 Max heart rate feature.

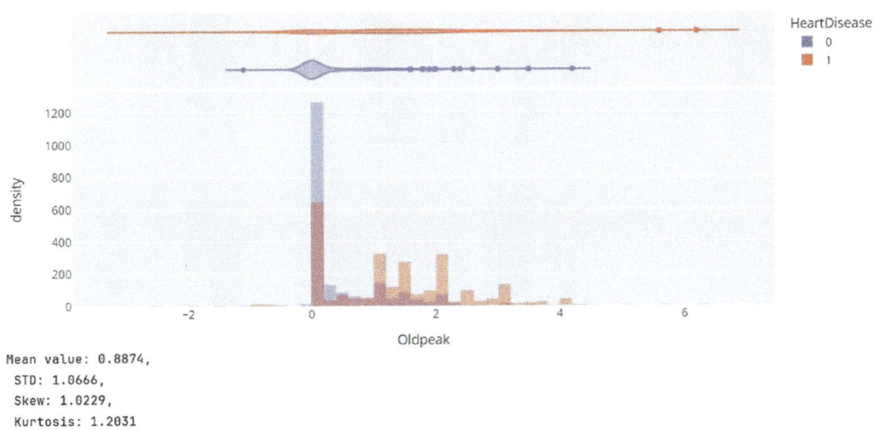

Figure 3.9 Oldpeak feature.

In the previous Figs. 3.5—3.9, we identified the dependence of quantitative features on the target class. Now let's do the same for qualitative features.

The following conclusions can be drawn from these six charts (Fig. 3.10).

- There are much more men with heart problems than women. About a third of women surveyed have heart problems, while only less than half of men are healthy.
- Almost all patients with asymptomatic chest pain have heart problems. In contrast, the rates for NAP and TA are almost equal, and people with atypical chest pain usually do not have heart problems.
- The absence of high blood sugar has almost no effect on the development of heart problems. On the contrary, its presence significantly increases the chances of developing heart disease.
- The best indicator of a heart problem on an electrogram is ST segment problems, while normal results and LV problems may indicate something else.
- The presence of angina during exercise can clearly indicate heart problems. After all, almost all patients with it suffer from heart disease.
- The flat and downward slope of the ST segment are clear indicators of heart disease. However, not the upward one, as most patients who have it do not experience heart problems.

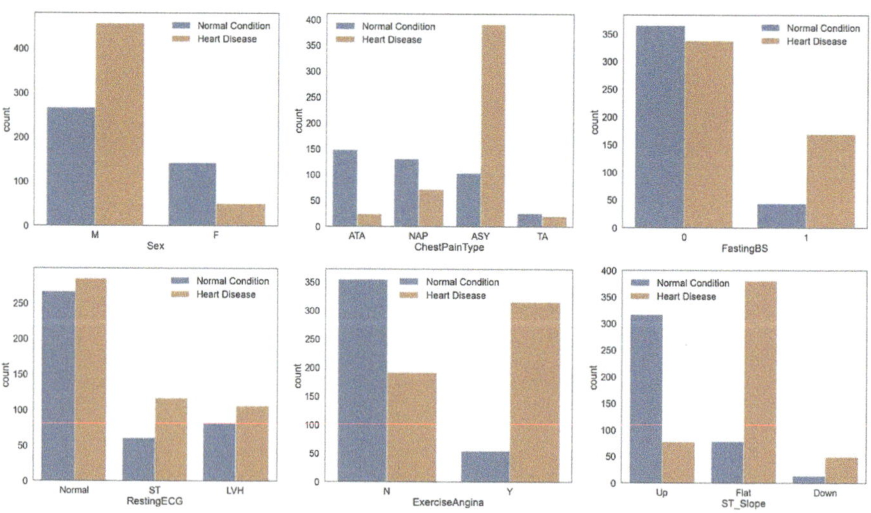

Figure 3.10 Qualitative features.

Target class

Now we should consider the target class. To do this, let's draw a pie chart (Fig. 3.11), with red representing patients with heart problems and yellow representing people with normal conditions. So, we see that the red sector accounts for 55.3%, which is $0.553 \times 918 = 508$ people. When the yellow sector has the remaining 44.7%, which in turn is equal to $0.447 \times 918 = 410$ patients (Klosterman, 2021).

This chart also shows that the data is evenly distributed and balanced. Therefore, there is no need to rebalance and create new records.

Correlation between features

If a healthcare professional knows what signs are important when assessing heart disease, it will help them make the right decision. This will also help not only doctors but also patients, as it will allow them to explain how the diagnosis was made and what was used to make it.

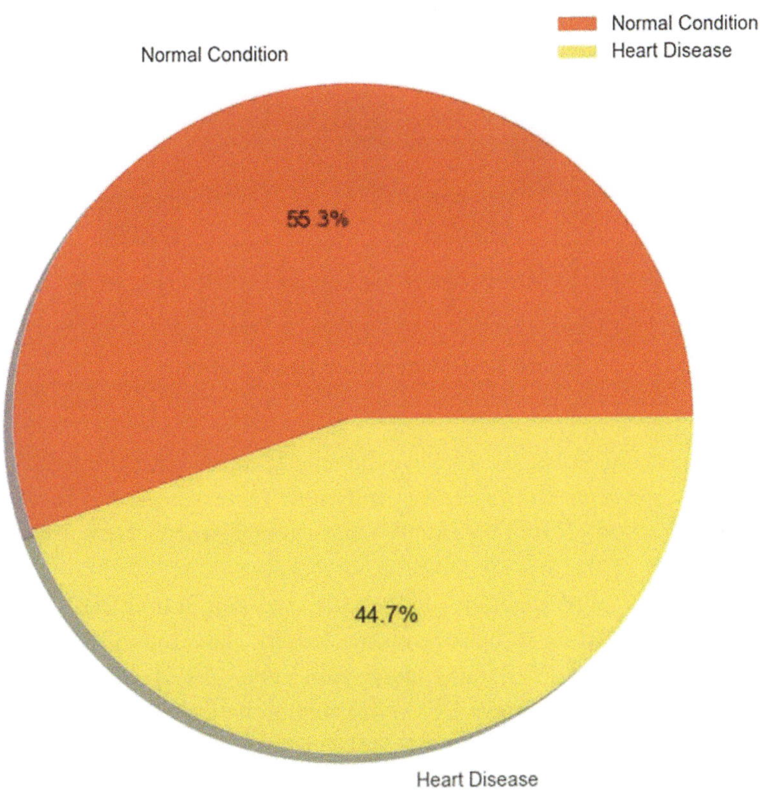

Figure 3.11 Heart disease distribution.

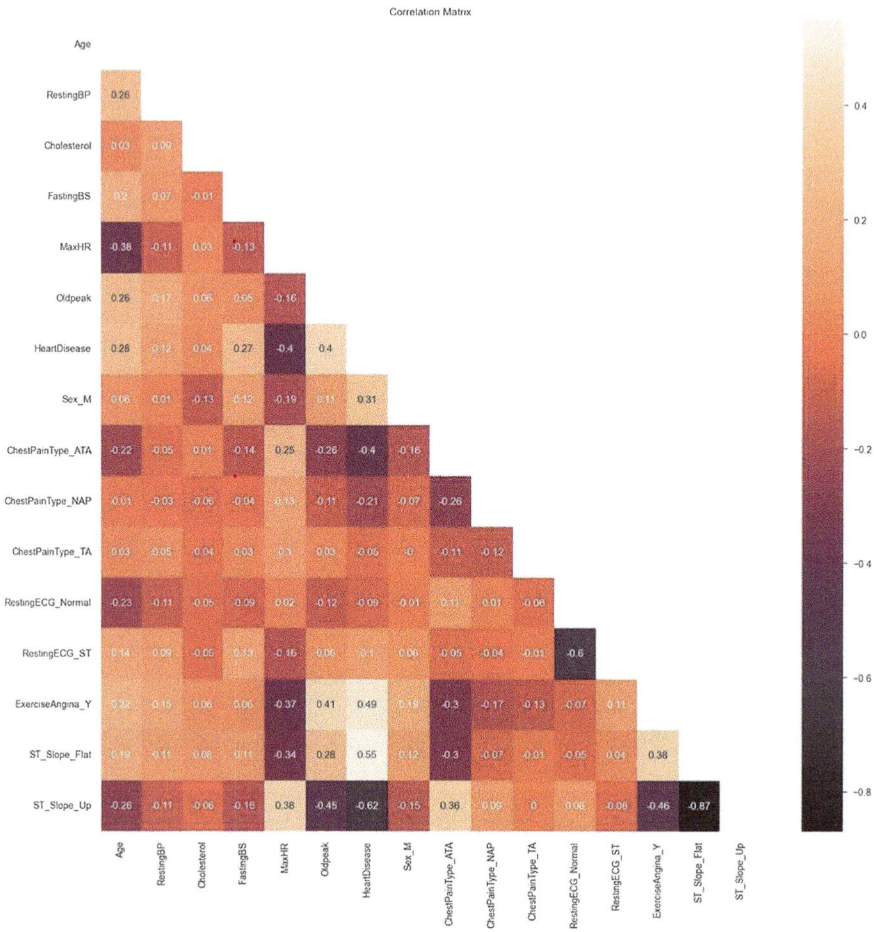

Figure 3.12 Feature correlation matrix.

Looking at Fig. 3.12, we can see that the features with the lightest and the darkest areas have the highest correlation. Here we should focusat the colors of the target Heart Disease class and other features. Now let's group the correlation of features depending on the class under study.

From this type of diagram (Fig. 3.13), we can immediately see that the highest correlation is between the desired class and ST Slope (Up, Flat), exercise angina (Yes), and chest pain type (ATA). And the lowest correlation is for the resting ECG (normal), chest pain type (TA), and cholesterol level.

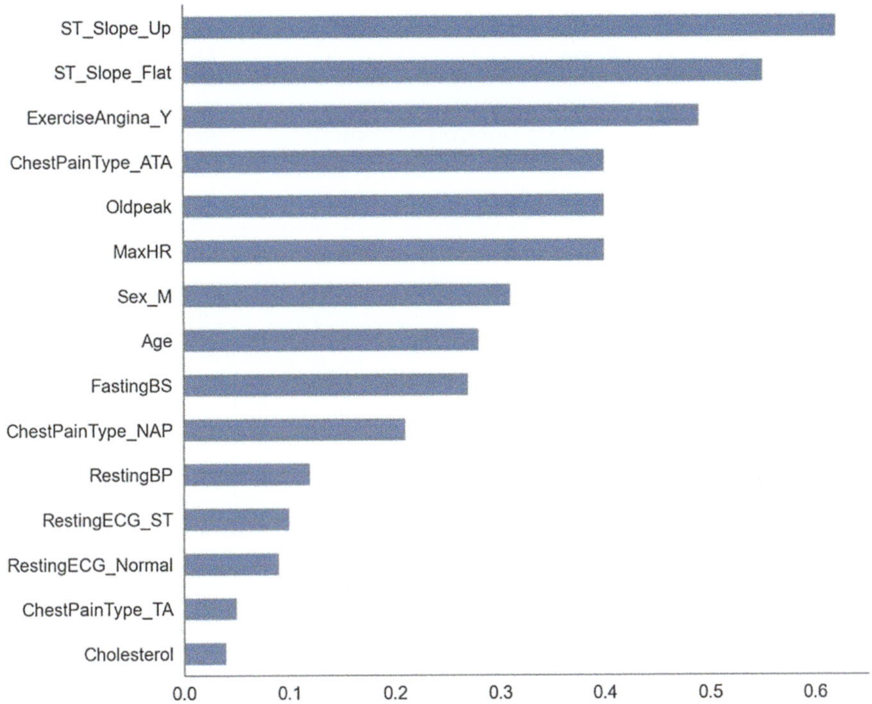

Figure 3.13 Correlation between features and target class.

Comparison of AI and XAI methods for classification task in healthcare

Types of models

Sixteen types of models were used to solve the problem of classifying the occurrence of a heart problem. They include Random Forest Classifier, Gradient Boosting Classifier, Extreme Gradient Boosting, Logistic Regression, Ridge Classifier, Linear Discriminant Analysis, Ada Boost Classifier, Naive Bayes, Spanning Tree Classifier, Light Gradient Boosting Machine, Quadratic Discriminant Analysis, Decision Tree Classifier K-neighbor classifier, SVM – linear kernel, dummy classifier and neural network ("Comparison of Gradient Boosting and Extreme Boosting Ensemble Methods for Webpage Classification," 2020; Sona et al., 2021; "Comparative analysis between the traditional K-Nearest Neighbor and Modifications with Weight-Calculation," 2022; What is a Decision Tree | IBM. IBM - Deutschland | IBM, 2023; Gao et al., 2016).

Table 3.7 Model types.

№	Name	Abbreviation
1	Random Forest Classifier	rf
2	Gradient Boosting Classifier	gbc
3	Extreme Gradient Boosting	xgboost
4	Logistic Regression	lr
5	Ridge Classifier	ridge
6	Linear Discriminant Analysis	lda
7	Ada Boost Classifier	ada
8	Naive Bayes	nb
9	Extra Trees Classifier	et
10	Light Gradient Boosting Machine	lightgbm
11	Quadratic Discriminant Analysis	qda
12	Decision Tree Classifier	dt
13	K Neighbors Classifier	knn
14	SVM – Linear Kernel	svm
15	Dummy Classifier	dummy
16	Neural Network	nn

As can be seen from Table 3.7. A total of 15 classifiers and one neural network were used. More detailed information about each model can be found at the reference column of the table.

Model creation

After the dataset has been processed, formatted, and transformed, it's time to create a model (Wang et al., 2023). The Pycaret library offers many possibilities for this. This can be done in the following method (Fig. 3.14).

As a result Table 3.8, the model goes through 10 training steps, and then the average value of the training metrics is selected (Jiyang, 2023). These metrics include: Accuracy, AUC, Recall, Precision, F1, Kappa, and MCC.

Another step is to build a confusion matrix. Fig. 3.15 shows the following results: the X – axis represents the class prediction results, and the Y – axis represents the true class label of the object (Doroshenko, 2019). Therefore, there are two types of classification errors: False Negative (FN) and False Positive (FP) (Díaz-Rodríguez et al., 2023).

```
rf = create_model('rf')
```

Executed at 2023.10.22 20:23:20 in 3s 297ms

Figure 3.14 Model creation.

Table 3.8 Random forest model.

Fold	Accuracy	AUC	Recall	Prec.	F1	Kappa	MCC
1	0.877	0.951	0.914	0.865	0.889	0.751	0.753
2	0.923	0.944	0.943	0.917	0.930	0.845	0.845
3	0.859	0.891	0.941	0.821	0.877	0.715	0.724
4	0.812	0.858	0.882	0.789	0.833	0.621	0.626
5	0.875	0.932	0.912	0.861	0.886	0.748	0.750
6	0.844	0.914	0.824	0.875	0.849	0.688	0.689
7	0.844	0.930	0.886	0.838	0.861	0.683	0.684
8	0.859	0.936	0.857	0.882	0.870	0.717	0.717
9	0.953	0.980	1.000	0.921	0.959	0.905	0.909
10	0.859	0.926	0.886	0.861	0.873	0.715	0.716
Mean	**0.871**	**0.926**	**0.904**	**0.863**	**0.883**	**0.739**	**0.741**
STD	**0.038**	**0.032**	**0.047**	**0.038**	**0.035**	**0.078**	**0.077**

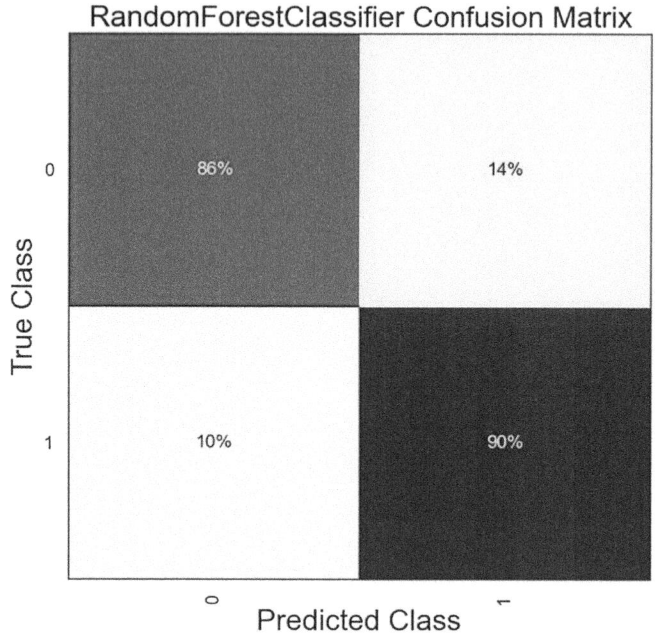

Figure 3.15 Confusion matrix for random forest.

The following metrics can be calculated from these errors.

- F1 score. Also called F − score or F − measure. In other words, F1 indicates the balance between precision and recall.

$$F_1 = \frac{Precision^{*} Recall}{Precision + Recall}$$

- Precision is the number True Positive predictions divided by the number of True Positive and False Positive. In other words, it is the number of positive predictions divided by the total number of predicted positive class values. It is also called positive predictive value (PPV).

$$Precision = \frac{TP}{TP + FP}$$

- Accuracy is the correctness rate of the classifier. Calculated with the help of the next formula:

$$Accuracy = \frac{TP + TN}{TP + TN + FP + FN}$$

- The Recall is the number of TPs divided by the number of TPs plus the number of FNs. It is also called the sensitivity or real TP value:

$$Recall = \frac{TP}{TP + FN}$$

Neural network

The following type of neural network was used to solve this problem (Fig. 3.16).

As we can see from the previous figure, the network is sequential and has four layers.

- One input with 11 neurons for each independent feature (age, sex, chest pain type, resting BP, cholesterol, fasting BS, resting ECG, maximum HR, exercise angina, Oldpeak, and ST slope).
- Two hidden layers with 128 neurons each.
- And since we have a binary classification, we have one output layer with one neuron for one class.

The training is performed in 50 epochs using the loss function – binary cross-tropy, Nadami optimizer, and the callback function EarlyStopping ("Waste Classifier using Naive Bayes Algorithm," 2022; Rojo-Álvarez et al., 2018; Luo and Liu, 2017).

The results of the classification can be seen in Fig. 3.17.

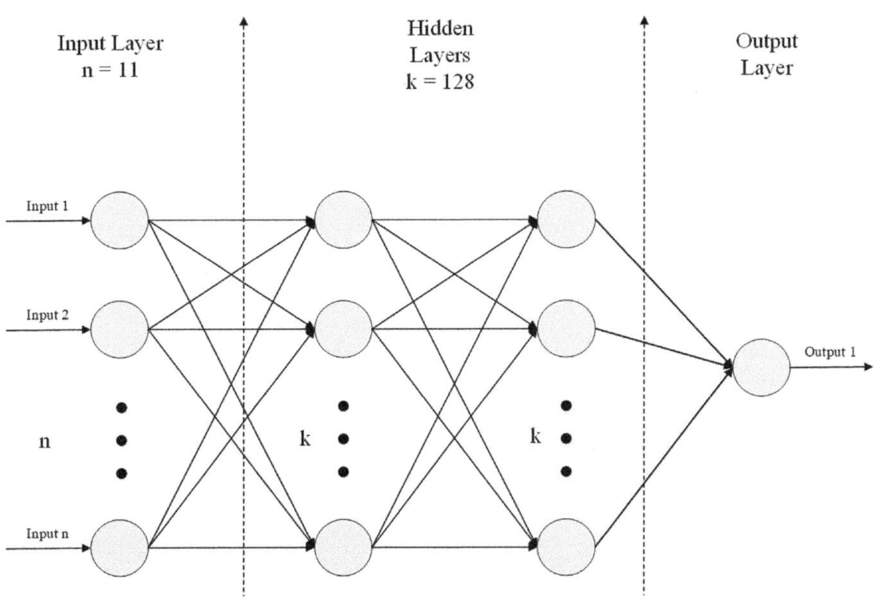

Figure 3.16 Neural network model.

```
a, score = model.evaluate(X_test, y_test)

print("Loss: " + str(a))

print("Accuracy: " + str(score))
Executed at 2023.10.11 15:07:36 in 172ms

8/8 [==============================] - 0s 2ms/step - loss: 2.5093 - accuracy: 0.8589
Loss: 2.5093259811401367
Accuracy: 0.8588709831237793
```

Figure 3.17 Results of NN classification.

As we can see above, the average classification accuracy is about 86%. This is about 1% less than other classifiers (Tezcan, 2021; Ghosh and Shuvo, 2019; Pekalska and Haasdonk, 2009).

Model comparison

Table 3.9 below shows that the best classifiers for the task are:

- Random Forest Classifier – Accuracy = 0.871 (87.1%), AUC = 0.926, Recall = 0.904, Prec. = 0.863, F1 = 0.883, Kappa = 0.739, MCC = 0.741, TT (Sec) = 0.191.
- Gradient Boosting Classifier – Accuracy = 0.868 (86.8%), AUC = 0.926, Recall = 0.893, Prec. = 0.866, F1 = 0.879, Kappa = 0.733, MCC = 0.734, TT (Sec) = 0.118.
- Extreme Gradient Boosting – Accuracy = 0.86 (86%), AUC = 0.919, Recall = 0.878, Prec. = 0.865, F1 = 0.871, Kappa = 0.718, MCC = 0.719, TT (Sec) = 0.114.
- Logistic Regression – Accuracy = 0.86 (86%), AUC = 0.925, Recall = 0.887, Prec. = 0.86, F1 = 0.872, Kappa = 0.717, MCC = 0.72, TT (Sec) = 0.194.

  And the worst are.
- K Neighbors Classifier – Accuracy = 0.674 (67.4%), AUC = 0.736, Recall = 0.737, Prec. = 0.685, F1 = 0.709, Kappa = 0.341, MCC = 0.343, TT (Sec) = 0.017.
- SVM – Linear Kernel – Accuracy = 0.566 (56.6%), AUC = 0, Recall = 0.445, Prec. = 0.637, F1 = 0.392, Kappa = 0.156, MCC = 0.214, TT (Sec) = 0.012.
- Dummy Classifier – Accuracy = 0.539 (53.9%), AUC = 0.5, Recall = 1, Prec. = 0.539, F1 = 0.7, Kappa = 0, MCC = 0, TT (Sec) = 0.007.

Table 3.9 Model results comparison.

Abbr	Accuracy	AUC	Rec.	Prec.	F1	Kap.	MCC	TT (sec)
Rf	0.871	0.926	0.904	0.863	0.883	0.739	0.741	0.191
Gbc	0.868	0.926	0.893	0.866	0.879	0.733	0.734	0.118
Xgb	0.860	0.919	0.878	0.865	0.871	0.718	0.719	0.114
Lr	0.860	0.925	0.887	0.860	0.872	0.717	0.720	0.194
Ridge	0.857	0.000	0.884	0.856	0.869	0.711	0.714	0.010
Lda	0.857	0.925	0.884	0.856	0.869	0.711	0.714	0.012
Ada	0.852	0.906	0.873	0.857	0.863	0.702	0.705	0.097
Nb	0.849	0.915	0.867	0.856	0.860	0.696	0.697	0.012
Et	0.847	0.917	0.875	0.848	0.861	0.692	0.694	0.165
Lgbm	0.846	0.925	0.867	0.851	0.858	0.689	0.691	0.049
Qda	0.836	0.901	0.861	0.841	0.850	0.670	0.672	0.011
Dt	0.763	0.763	0.772	0.790	0.778	0.524	0.529	0.013
Knn	0.674	0.736	0.737	0.685	0.709	0.341	0.343	0.017
Svm	0.566	0.000	0.445	0.637	0.392	0.156	0.214	0.012
Dum	0.539	0.500	1.000	0.539	0.700	0.000	0.000	0.007

Explaining with permutation feature importance and SHAP

After building the models, it is necessary to analyze the results obtained, for which it is advisable to build graphs of the importance of the features (Sokol and Flach, 2020). In total, two types of feature importance were studied: Permutation feature importance and SHAP feature importance. The use of these methods makes it possible to increase the explainability of applied AI methods (Lipton, 2018).

Permutation feature importance — determines the amount of reduction in the model score if the values of one feature are randomly shuffled. This type of method tries to break the relationship between the feature and the target class to check the level of model score drop (Hoffman et al., 2018). The more it drops, the more important the feature is.

From Fig. 3.18 below, you can see that the most important characteristics in assessing the presence of heart disease in a patient are:

- **Maximum heart rate.** So, the lower it is, the more likely a person is to have heart problems. The same conclusion can be reached by looking at
- **Fasting blood sugar** is a clear indicator of heart disease. The higher it is, the more likely a person will have problems in the future.
- If you are a **male**, the probability of heart disease increases significantly, so men need to have regular heart examinations after the age of 50.
- If your **resting ECG** shows **normal** results, this does not mean that you are fine. This may indicate the presence of other hidden risks.
- If the patient has **atypical angina**, this indicates a violation of the heart and can lead to the appearance of various heart diseases.

And the characteristics of lesser importance are the following:

- The **upward slope** of the **ST.** Usually indicates normal heart behavior. This can be clearly seen in, as almost all patients with this type of slope have no heart problems.
- If a **resting electrocardiogram** shows abnormalities with **ST** elevation, this usually does not mean a problem with heart.

Figure 3.18 Permutation feature importance for random forest.

- **Angina during exercise** is not a good indicator of heart disease. It may indicate any other types of disease that can be detected after a visit to a doctor.
- Since the **age** of the surveyed patients was approximately the same and exceeded 50 years, this is not a significant indicator of health problems in this option.
- **Typical angina** does not greatly affect the results of the classification of heart problems. This may indicate both a low diversity of patients and other types of diseases that cause this score.

The next option for the importance of features is SHAP (Chaddad et al., 2023). This type of chart shows which feature has the most influence on the prediction of one observation relative to other features in that observation and the mean result across the data set (Fauvel et al., 2020).

In Fig. 3.19, positive SHAP values indicate higher values of the feature, while negative SHAP values indicate lower values of the feature. The

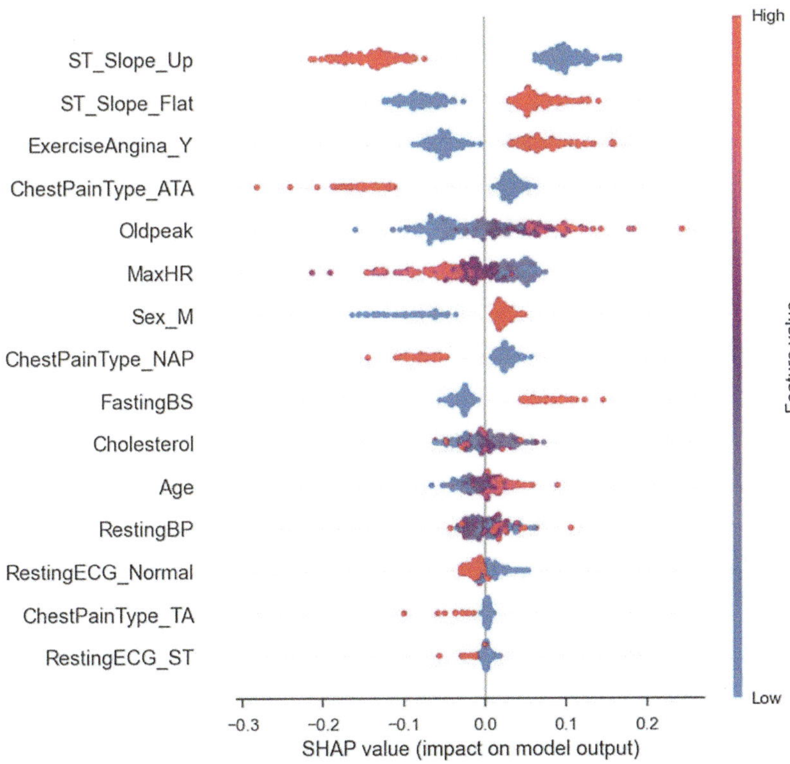

Figure 3.19 SHAP feature importance for random forest.

legend shows that higher values are marked in red and lower values in blue. This figure also shows that the most important features are upward and flat ST slope, exercise angina – yes, chest pain type ATA and Oldpeak feature (Larochelle et al., 2007). Less important are the following: resting blood pressure, resting ECG – normal, type of chest pain – typical angina, and resting ECG – ST.

If we check out this diagram for a neural network (Fig. 3.20), we see the identical picture. The same important features are true for this type of model as for a Random Forest.

Recursive feature selection

Another good technique is recursive feature elimination (RFE) (Goodfellow et al., 2020). This method identifies the weakest features in the model and removes them until the desired number of features is reached (Huynh et al., 2022).

In this Fig. 3.21, we can see that a total of 14 characteristics were selected, but after 12, the efficiency score almost does not change. This may indicate that a larger number of features is not always a good thing. And that the initial number of symptoms (76) in the dataset was clearly

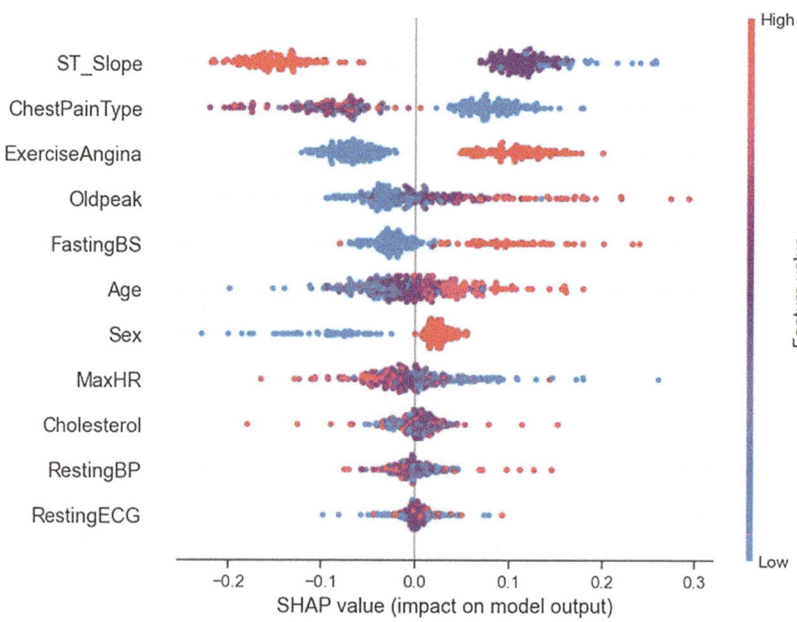

Figure 3.20 SHAP feature importance for neural network.

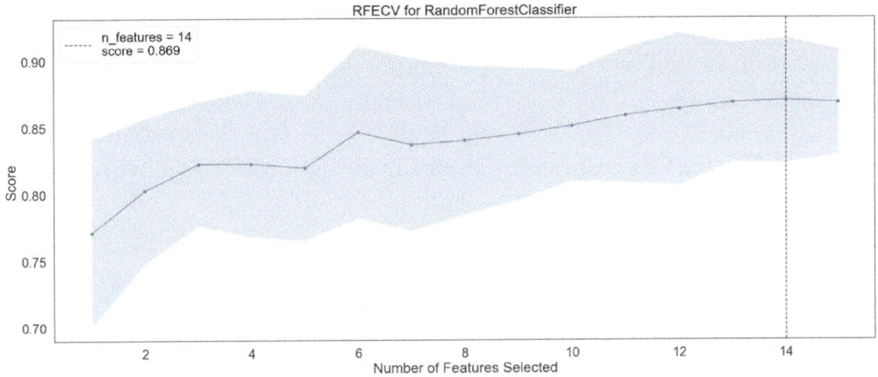

Figure 3.21 Recursive feature elimination.

redundant (Sarkar, 2022). Therefore, it is not always advisable for healthcare professionals to ask patients about all possible symptoms, but to select only a certain number that have the greatest impact on the assessment results.

Decision boundary

Another good method is to plot the decision boundary or, in other words, the decision surface (Velden et al., 2022). By its very essence, this type of diagram displays the surface in the form of two subsets represented by different colors, one for each value of the target class. Thus, the model classifies the drawn surface according to the label of the desired class.

As can be seen from Fig. 3.22, the objects of the desired class are not explicitly separated and have many outliers. This leads to a decrease in the quality of the classification and problems in building the model (Simonyan et al., 2019).

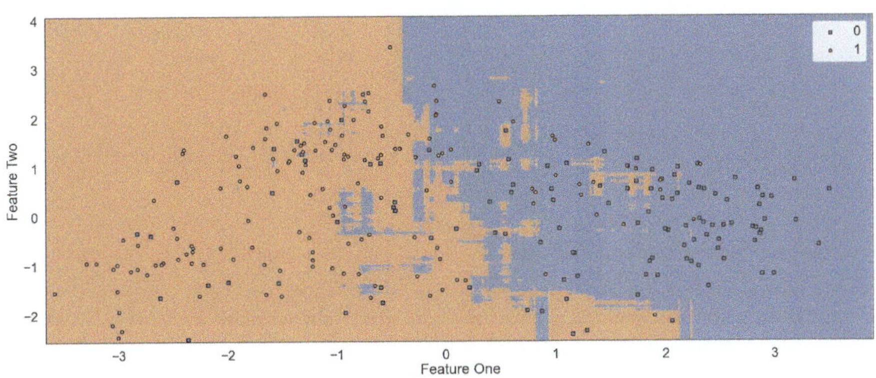

Figure 3.22 Decision boundary of the target class.

Explaining with decision tree machine learning model

The Python toolkit allows you to automatically generate decision trees for tree-based classifiers (Extra Trees, Decision Tree, Random Forest, etc.) (Amann et al., 2020). This will help explain to healthcare professionals and patients how the decision about their disease was made and what factors preceded it (Manresa-Yee et al., 2022).

In essence, a decision tree (Fig. 3.23) consists of a single node that then branches into different possible outcomes (Khodabandehloo et al., 2021). Each of these outcomes leads to certain additional nodes, which, in turn, branch to other nodes. This is how a tree structure is formed. In general, there are three types of nodes: chance nodes, decision nodes, and end nodes (Joyce et al., 2023).

As can be seen from Fig. 3.24, the node was first divided by maximum Heart Rate $\leq$ 96.5, and then the ST Slope Up characteristic $\leq$ 0.5 was checked. This process continues until all the final nodes (target class label) are found.

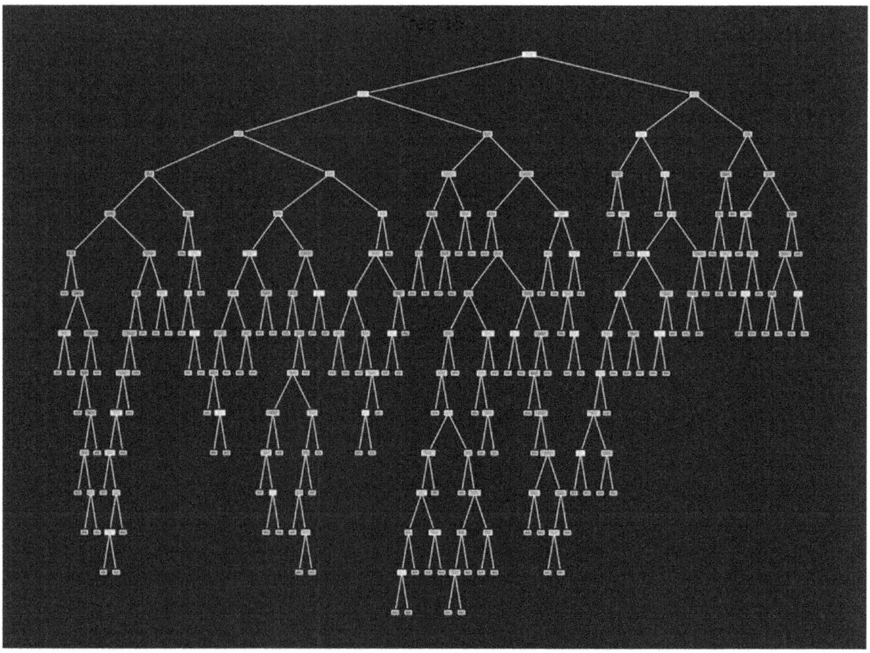

Figure 3.23 One of the decision trees for random forest.

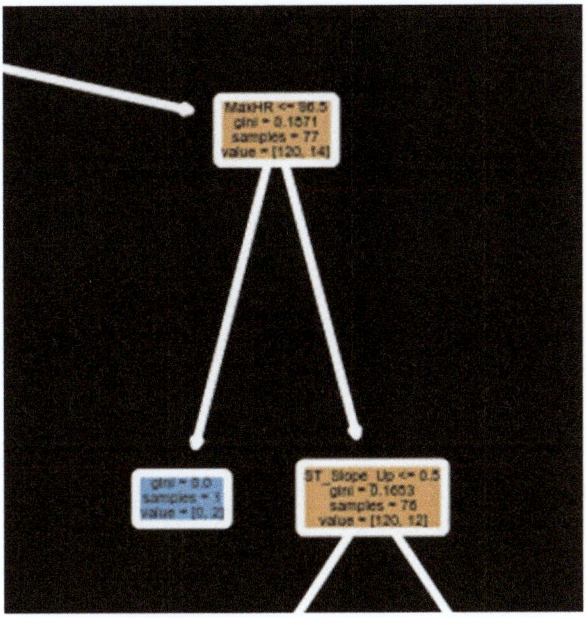

Figure 3.24 Set of decision nodes.

Conclusions

The study analyzed the existing legislative acts regulating the use of artificial intelligence systems. It was determined that the main requirements of the AI Act in particular are the transparency of the use of AI systems, the legality of obtaining training data, and their accuracy (Asan et al., 2020). It is important to carefully discover the training samples before using them in AI systems to avoid the possibility of discrimination or bias in the trained systems. An example of such preprocessing of data for the medical task of identifying heart diseases is given in the study.

It is also important for the medical field to maximize the explainability of the methods used in order to increase the trust of users (both doctors and patients) in the decisions made by the AI system. In this chapter, various machine learning models that can help detect and treat cardiovascular disease at an early stage are explored and developed. These models include Random Forest Classifier, Gradient Boosting Classifier, Extreme Gradient Boosting, Logistic Regression, Ridge Classifier, Linear Discriminant Analysis, Ada Boost Classifier, Naive Bayes, spanning tree classifier, light gradient boosting machine, quadratic discriminant analysis, decision tree classifier, K-neighbor classifier, SVM – support vector machine, Dummy classifier and Neural Network. The best results were shown by Random Forest – 87%, Gradient Boosting – 86.8%, Logistic Regression – 86%, and Neural Network 86%. Analyzing the obtained results, we defined that several methods gave almost the same classification accuracy. In such a case, it is desirable to choose a method that is more explainable. In this case, it is the Random Forest method.

Permutation feature importance and SHAP are additionally useful methods that are convenient to use for the analysis of features and their influence on the decision made, even when using methods belonging to the black box methods. These methods show which input features had the greatest impact on the decision made and were of decisive importance.

Acknowledgments

This work was realized within the framework of the program Erasmus+ Jean Monnet Module "Trustworthy artificial intelligence: the European approach" (101085626 – TrustAI – ERASMUS- JMO-2022-HEI-TCH-RSCH).

References

Alizadehsani, R., Roshanzamir, M., Abdar, M., Beykikhoshk, A., Khosravi, A., Panahiazar, M., Koohestani, A., Khozeimeh, F., Nahavandi, S., Sarrafzadegan, N., 2019. A database for using machine learning and data mining techniques for coronary artery disease diagnosis. Sci. Data 6 (1). https://doi.org/10.1038/s41597-019-0206-3.

Amann, J., Blasimme, A., Vayena, E., Frey, D., Madai, V.I., 2020. Explainability for artificial intelligence in healthcare: a multidisciplinary perspective. BMC Med. Inf. Decis. Making 20 (1), 310. https://doi.org/10.1186/s12911-020-01332-6.

Asan, O., Bayrak, A.E., Choudhury, A., 2020. Artificial intelligence and human trust in healthcare: focus on clinicians. J. Med. Internet Res. 22 (6). https://doi.org/10.2196/15154.

Barredo Arrieta, A., Díaz-Rodríguez, N., Del Ser, J., Bennetot, A., Tabik, S., Barbado, A., Garcia, S., Gil-Lopez, S., Molina, D., Benjamins, R., Chatila, R., Herrera, F., 2020. Explainable Artificial Intelligence (XAI): concepts, taxonomies, opportunities and challenges toward responsible AI. Inf. Fusion 58, 82–115. https://doi.org/10.1016/j.inffus.2019.12.012.

Bharati, S., Mondal, M.R.H., Podder, P., 2023. A review on explainable artificial intelligence for healthcare: why, how, and when? IEEE Transactions on Artificial Intelligence 1–15. https://doi.org/10.1109/tai.2023.3266418.

Chaddad, A., Peng, J., Xu, J., Bouridane, A., 2023. Survey of explainable AI techniques in healthcare. Sensors 23 (2). https://doi.org/10.3390/s23020634.

Comparative analysis between the traditional K-nearest neighbor and modifications with weight-calculation. In: ISMSIT 2022 - 6th International Symposium on Multidisciplinary Studies and Innovative Technologies, Proceedings, 2022, pp. 961–965. https://doi.org/10.1109/ISMSIT56059.2022.9932693.

Comparison of gradient boosting and extreme boosting ensemble methods for webpage classification. In: Proceedings - 2020 5th International Conference on Research in Computational Intelligence and Communication Networks, ICRCICN 2020, 2020, pp. 77–82. https://doi.org/10.1109/ICRCICN50933.2020.9296176.
Currie, G., Hawk, K.E., Rohren, E., Vial, A., Klein, R., 2019. Machine learning and deep learning in medical imaging: intelligent imaging. J. Med. Imag. Radiat. Sci. 50 (4), 477–487. https://doi.org/10.1016/j.jmir.2019.09.005.
Díaz-Rodríguez, N., Del Ser, J., Coeckelbergh, M., López de Prado, M., Herrera-Viedma, E., Herrera, F., 2023. Connecting the dots in trustworthy Artificial Intelligence: from AI principles, ethics, and key requirements to responsible AI systems and regulation. Inf. Fusion 99. https://doi.org/10.1016/j.inffus.2023.101896.
Doroshenko, A., 2019. Application of global optimization methods to increase the accuracy of classification in the data mining tasks. Computer Modeling and Intelligent Systems 2353, 98–109. https://doi.org/10.32782/cmis/2353-8.
Doroshenko, A., 2022. Trustworthy AI in medicine and healthcare - abstract. In: International Workshop on Informatics & Data- Driven Medicine.
Fauvel, K., Masson, V., Fromont, E., 2020. A Performance-Explainability Framework to Benchmark Machine Learning Methods: Application to Multivariate Time Series Classifiers. https://doi.org/10.48550/arXiv.2005.14501.
Gao, Y., Rong, W., Shen, Y., Xiong, Z., 2016. Convolutional Neural Network based sentiment analysis using Adaboost combination. In: Proceedings of the International Joint Conference on Neural Networks, pp. 1333–1338. https://doi.org/10.1109/IJCNN.2016.7727352.
Gerlings, J., Jensen, M.S., Shollo, A., 2022. Explainable AI, but explainable to whom? An exploratory case study of xAI in healthcare. In: Intelligent Systems Reference Library, vol 212, pp. 169–198. https://doi.org/10.1007/978-3-030-83620-7_7.
Ghosh, J., Shuvo, S.B., 2019. Improving classification model's performance using linear discriminant analysis on linear data. In: 2019 10th International Conference on Computing, Communication and Networking Technologies, ICCCNT 2019. https://doi.org/10.1109/ICCCNT45670.2019.8944632.
Goodfellow, I., Pouget-Abadie, J., Mirza, M., Xu, B., Warde-Farley, D., Ozair, S., Courville, A., Bengio, Y., 2020. Generative adversarial networks. Commun. ACM 63 (11), 139–144. https://doi.org/10.1145/3422622.
Goyal, L., Dhull, A., Singh, A., Kukreja, S., Singh, K.K., 2022. VGG-COVIDNet: a Novel model for COVID detection from X-Ray and CT Scan images. Proc. Comput. Sci. 218, 1926–1935. https://doi.org/10.1016/j.procs.2023.01.169.
Haresamudram, K., Larsson, S., Heintz, F., 2023. Three levels of AI transparency. Computer 56 (2), 93–100. https://doi.org/10.1109/MC.2022.3213181.
Heart Failure Prediction Dataset, 2021. https://www.kaggle.com/datasets/fedesoriano/heart-failure-prediction. (Accessed 10 October 2023).
Hoffman, R.R., Mueller, S.T., Klein, G., Litman, J., 2018. Metrics for explainable AI: challenges and prospects. arXiv. https://doi.org/10.48550/arxiv.1812.04608.
Hohma, E., Lütge, C., 2023. From trustworthy principles to a trustworthy development process: the need and elements of trusted development of AI systems. AI (Switzerland) 4 (4), 904–925. https://doi.org/10.3390/ai4040046.
How to Use AI and Personal Data Appropriately and Lawfully, 2022, 2023. Information Commissioner's Office (ICO).
Hulsen, T., 2023. Explainable artificial intelligence (XAI): concepts and challenges in healthcare. AI 4 (3), 652–666. https://doi.org/10.3390/ai4030034.
Huynh, T.D., Tsakalakis, N., Helal, A., Stalla-Bourdillon, S., Moreau, L., 2022. Explainability-by-Design: A Methodology to Support Explanations in Decision-Making Systems. https://doi.org/10.48550/arXiv.2206.06251.

Jangra, M., Dhull, S.K., Singh, K.K., Singh, A., 2023. Denoising techniques for ECG arrhythmia classification systems: an experimental approach. Lect. Notes Data Eng. Commun. Technol. 175, 1–13. https://doi.org/10.1007/978-981-99-3432-4_1.

Jiyang, P., 2023. Research of classification methods based on machine learning. In: 2023 2nd International Conference on Health Big Data and Intelligent Healthcare (ICHIH), Zhuhai, China. pp. 63–67. https://doi.org/10.1109/ICHIH60370.2023.10396622.

Joyce, D.W., Kormilitzin, A., Smith, K.A., Cipriani, A., 2023. Explainable artificial intelligence for mental health through transparency and interpretability for understandability. npj Digit. Med. 6 (1). https://doi.org/10.1038/s41746-023-00751-9.

Jung, J., Lee, H., Jung, H., Kim, H., 2023. Essential properties and explanation effectiveness of explainable artificial intelligence in healthcare: a systematic review. Heliyon 1611.

Khodabandehloo, E., Riboni, D., Alimohammadi, A., 2021. HealthXAI: collaborative and explainable AI for supporting early diagnosis of cognitive decline. Future Generat. Comput. Syst. 116, 168–189. https://doi.org/10.1016/j.future.2020.10.030.

Khurana, R., Choudhary, M., Singh, A., Singh, K.K., 2024. Energy-efficient fog-assisted system for monitoring diabetic patients with cardiovascular disease. In: Blockchain and Deep Learning for Smart Healthcare, pp. 323–352. https://www.scopus.com/inward/record.uri?eid=2-s2.0-85180870016&partnerID=40&md5=e8a81dafe2fb34af69c2670ce843ab28.

Klosterman, S., 2021. Data Science Projects with Python: A Case Study Approach to Gaining Valuable Insights from Real Data with Machine Learning. Packt Publishing. http://ieeexplore.ieee.org/document/10162389.

Larochelle, H., Erhan, D., Courville, A., Bergstra, J., Bengio, Y., 2007. An empirical evaluation of deep architectures on problems with many factors of variation. ACM Int. Conf. Proceeding Ser. 227, 473–480. https://doi.org/10.1145/1273496.1273556.

Lipton, Z.C., 2018. The mythos of model interpretability: in machine learning, the concept of interpretability is both important and slippery. Queue 16 (3). http://dl.acm.org/citation.cfm?id=J882.

Loyola-Gonzalez, O., 2019. Black-box vs. white-box: understanding their advantages and weaknesses from a practical point of view. IEEE Access 7, 154096–154113. https://doi.org/10.1109/access.2019.2949286.

Luo, H., Liu, Y., 2017. A prediction method based on improved ridge regression. In: Proceedings of the IEEE International Conference on Software Engineering and Service Sciences, ICSESS, pp. 596–599. https://doi.org/10.1109/ICSESS.2017.8342986.

Manresa-Yee, C., Roig-Maimó, M.F., Ramis, S., Mas-Sansó, R., 2022. Advances in XAI: explanation interfaces in healthcare. In: Intelligent Systems Reference Library, vol 212, pp. 357–369. https://doi.org/10.1007/978-3-030-83620-7_15.

Pekalska, E., Haasdonk, B., 2009. Kernel discriminant analysis for positive definite and indefinite kernels. IEEE Trans. Pattern Anal. Mach. Intell. 31 (6), 1017–1032. https://doi.org/10.1109/tpami.2008.290.

Proposal for a regulation of the European parliament and of the council on harmonised rules on fair access to and use of data (Data Act). European Union.

Public Voice Coalition, Universal Guidelines for Artificial Intelligence, 2018.

Rai, A., 2020. Explainable AI: from black box to glass box. J. Acad. Market. Sci. 48 (1), 137–141. https://doi.org/10.1007/s11747-019-00710-5.

Regulation (EU) 2022/868 of the European Parliament and of the Council of 30 May 2022 on European Data Governance and Amending Regulation (EU) 2018/1724 (Data Governance Act, 2022 (European Union).

Ribeiro, M.T., Singh, S., Guestrin, C., 2016. "Why should i trust you?" Explaining the predictions of any classifier. In: Proceedings of the ACM SIGKDD International Conference on Knowledge Discovery and Data Mining, vol 13–17, pp. 1135–1144. https://doi.org/10.1145/2939672.2939778.

Rojo-Álvarez, J.L., Martínez-Ramón, M., Muñoz-Marí, J., Camps-Valls, G., 2018. Support vector machine and kernel classification algorithms. In: Digital Signal Processing with Kernel Methods, pp. 433–502. https://doi.org/10.1002/9781118705810.ch10.

Sarkar, A., 2022. Is Explainable AI a Race against Model Complexity? https://doi.org/10.48550/arXiv.2205.10119.

Savchuk, D., Doroshenko, A., 2021. Investigation of machine learning classification methods effectiveness. In: International Scientific and Technical Conference on Computer Sciences and Information Technologies, vol 1, pp. 33–37. https://doi.org/10.1109/CSIT52700.2021.9648582.

Shafay, M., Ahmad, R.W., Salah, K., Yaqoob, I., Jayaraman, R., Omar, M., 2023. Blockchain for deep learning: review and open challenges. Cluster Comput. 26 (1), 197–221. https://doi.org/10.1007/s10586-022-03582-7.

Sharma, U., Ganapathi, A., Singh, A., Singh, K.K., 2024. Blockchain in healthcare: use cases. In: Blockchain and Deep Learning for Smart Healthcare, pp. 147–169. https://www.scopus.com/inward/record.uri?eid=2-s2.0-85180832541&partnerID=40&md5=a909836d141bd968e3d48f2fd01c7c93.

Sheu, R.-K., Pardeshi, M.S., 2022. A survey on medical explainable AI (XAI): recent progress, explainability approach, human interaction and scoring system. Sensors. https://doi.org/10.3390/s22208068.

Simonyan, K., Vedaldi, A., Zisserman, A., 2019. Deep Inside Convolutional Networks: Visualising Image Classification Models and Saliency Maps.

Singh, A., Kant Singh, K., 2024. Blockchain and deep learning: research challenges, open problems, and future. In: Blockchain and Deep Learning for Smart Healthcare, pp. 425–431. https://www.scopus.com/inward/record.uri?eid=2-s2.0-85180886190&partnerID=40&md5=77f9ee3b59ff54e89a75c28eccd90e39.

Sokol, K., Flach, P., 2020. Explainability fact sheets: a framework for systematic assessment of explainable approaches. In: FAT* 2020 - Proceedings of the 2020 Conference on Fairness, Accountability, and Transparency, pp. 56–67. https://doi.org/10.1145/3351095.3372870.

Sona, D.S., Asha, D.S., Samarjeet, B., 2021. 5. Assimilate Machine Learning Algorithms in Big Data Analytics: Review, pp. 81–114. http://ieeexplore.ieee.org/document/9478287.

Tezcan, B., 2021. Why using a dummy classifier is a smart move. (Accessed 12 October 2023).

Tjoa, E., Guan, C., 2021. A survey on explainable artificial intelligence (XAI): toward medical XAI. IEEE Transact. Neural Networks Learn. Syst. 32 (11), 4793–4813. https://doi.org/10.1109/tnnls.2020.3027314.

UCI Machine Learning Repository, 2023. https://archive.ics.uci.edu/dataset/45/heart+disease. (Accessed 10 October 2023).

Velden, Kuijf, H.J., Gilhuijs, K.G.A., Viergever, M.A., 2022. Explainable artificial intelligence (XAI) in deep learningbased medical image analysis. Med. Image Anal. 79, 102470. https://doi.org/10.1016/j.media.2022.102470.

Wang, Y.-C., Chen, T.-C.T., Chiu, M.-C., 2023. An improved explainable artificial intelligence tool in healthcare for hospital recommendation. Healthcare Anal. 3. https://doi.org/10.1016/j.health.2023.100147.

Waste classifier using naive Bayes algorithm. In: 2022 10th International Conference on Cyber and IT Service Management, CITSM 2022, 2022. https://doi.org/10.1109/CITSM56380.2022.9935894.

What is a Decision Tree | IBM. IBM - Deutschland | IBM, 2023. https://www.ibm.com/topics/decision-trees. (Accessed 10 October 2023).

Yang, G., Ye, Q., Xia, J., 2022. Unbox the black-box for the medical explainable AI via multi-modal and multi-centre data fusion: a mini-review, two showcases and beyond. Inf. Fusion 77, 29–52. https://doi.org/10.1016/j.inffus.2021.07.016.

CHAPTER 4

Designing transparent and accountable AI systems for healthcare

Tripti Sharma[1], Anuradha Dhull[1], Akansha Singh[2] and Krishna Kant Singh[3]

[1]Department of CSE, The NorthCap University, Gurugram, Haryana, India; [2]School of CSET, Bennett University, Greater Noida, Uttar Pradesh, India; [3]Delhi Technical Campus, Greater Noida, Uttar Pradesh, India

Introduction

The European Commission states that enhancing healthcare is one-way artificial intelligence (AI) will improve people's lives (European Commission, 2020, Alowais et al., 2023). AI has enormous potential to improve diagnosis and clinical care, health research and drug development, resource allocation efficiency, and healthcare management. AI is driven by the growing availability of healthcare data and the quick advancement of analytical techniques (Amann, 2022).

The purpose of this study is to investigate the significance of developing accountable and transparent AI systems for the healthcare industry. It will look at how AI is currently being used in healthcare, stressing both the advantages and difficulties of integrating AI. The paper will also go over tactics and best practices for encouraging accountability and openness in AI-driven healthcare apps. By analyzing case studies, ethical frameworks, and regulatory guidelines, this paper seeks to provide insights into how healthcare organizations can navigate the complexities of AI implementation while prioritizing patient safety, privacy, and trust. Ultimately, the goal is to contribute to the development of ethical, responsible, and beneficial AI systems that empower healthcare professionals and improve patient care. The rapid development of artificial intelligence technology has an important impact across many industries and societies (Drysdale, 2019).

Responsible and Explainable Artificial Intelligence in Healthcare
ISBN 978-0-443-24788-0
https://doi.org/10.1016/B978-0-443-24788-0.00004-2

Exploring artificial intelligence

AI was introduced in 1956, has soared to become a leading technology. Often described as mimicking the human brain, AI empowers machines to perceive, reason, and act like us. (Fig. 4.1 provides a deeper look at its evolution.)

This intelligence is transforming healthcare. AI promises to revolutionize patient care, diagnosis, and treatment, boosting efficiency, accuracy, and accessibility. It could even improve outcomes and cut costs (Ethics and governance of artificial intelligence for health: WHO guidance, 2021). However, ethical, legal, and social concerns accompany this exciting shift.Transparency and accountability are crucial to ensure patient safety, privacy, and trust in AI systems.

Key changes made are as follows:

- Removed specific references to the source material.
- Simplified technical terms "perception" and "decision-making".
- Replaced "leverage machines" with "empower machines".
- Shortened and simplified sentences.
- Added a clearer transition sentence.
- Focused on the key message: AI's potential in healthcare and the ethical consideration

Fig. 4.1 shows the evolution of AI over time.

AI and ML: Revolutionizing healthcare

Machine Learning (ML), a subset of Artificial Intelligence (AI), enables systems to learn from data without explicit programming. Algorithms analyze labeled data to identify patterns and make predictions or judgments, enhancing patient care, diagnosis, and treatment in healthcare. As shown in Fig. 4.2 we can get to know the road map from clinical data generation to data analysis ML facilitates early illness detection, precise diagnosis, and personalized treatment regimens. It also streamlines administrative tasks, accelerates drug discovery, and optimizes surgical procedures. Predictive analytics driven by AI forecast disease outbreaks and identified at-risk patients, while conversational AI interfaces enhance patient engagement. Despite challenges such as algorithmic bias and data privacy, AI and ML have the potential to

Figure 4.1 AI evolution AI evolution.

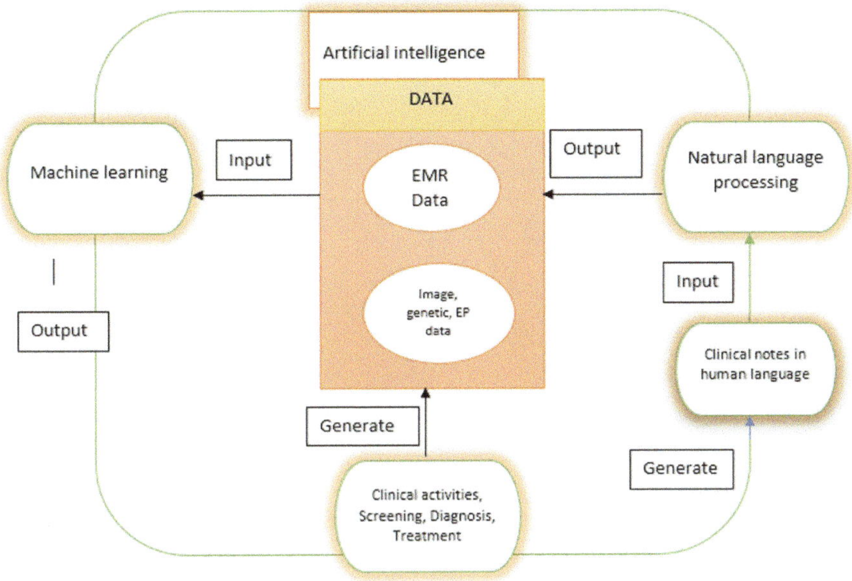

Figure 4.2 Creation path creation path.

transform healthcare into a proactive, predictive, and tailored system benefiting both patients and professionals (Felzmann et al., 2020).

Understanding the importance of transparency in healthcare AI

Transparency in AI healthcare systems refers to the visibility and comprehensibility of the inner workings, processes, and decision-making mechanisms of artificial intelligence algorithms utilized in healthcare settings. It entails making the algorithms, data sources, training methodologies, and decision outputs accessible and understandable to relevant stakeholders, including healthcare providers, patients, researchers, and policymakers. Transparent AI healthcare systems enable stakeholders to understand how AI algorithms arrive at their conclusions, including the factors considered, biases present, and uncertainties involved, which can be better understood by Fig. 4.3. This transparency fosters trust, accountability, and informed decision-making, ensuring that AI technologies are deployed ethically, responsibly, and effectively. By promoting visibility and clarity, transparency in AI healthcare systems refer empowers stakeholders to assess the reliability, fairness, and potential risks associated with AI applications, thus facilitating the advancement of patient-centered, equitable, and ethical healthcare practices (Elkhodr et al., 2023).

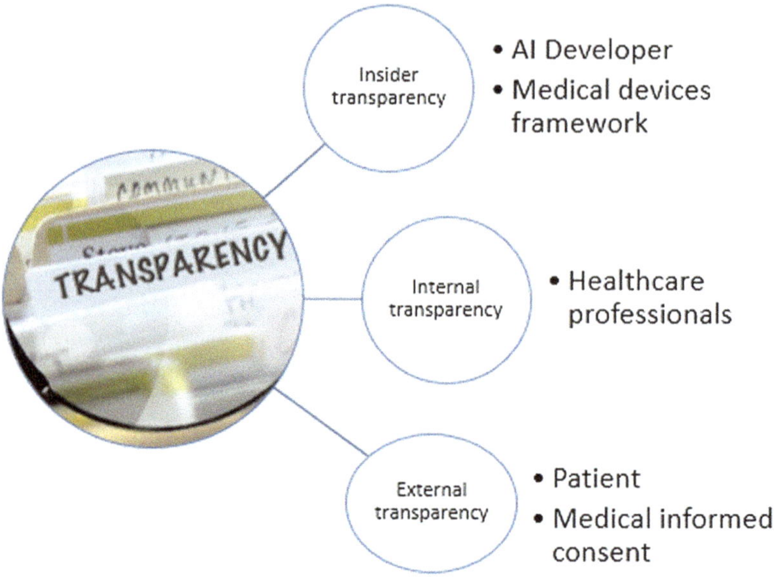

Figure 4.3 Healthcare AI's layered transparency system healthcare AI's layered transparency system.

Ethical considerations for AI in medical settings

Kiseleva et al. (2020) researched that we need to ensure the responsible and effective use of AI in medical settings, it is crucial to address ethical considerations such as accountability, responsibility, and trust. This includes examining the decision-making processes and potential biases of AI systems, as well as addressing concerns related to algorithmic bias, privacy, and data security. The development of transparent and accountable AI machine learning systems in healthcare is crucial to address the challenges and ethical considerations associated.

Challenges of implementing AI and ML in the medical field

One of the key challenges in implementing AI and ML in the medical field is ensuring transparency and accountability. Without transparency, it becomes difficult to understand how AI and ML algorithms are making decisions, leading to concerns about bias, discrimination, and ethical implications. Furthermore, accountability is crucial to ensure that AI and ML systems are held responsible for their actions and outcomes. To overcome these challenges, robust and prospective clinical evaluation is essential to ensure the safety and effectiveness of AI systems in healthcare.

Organizations must also prioritize ethical practices, addressing issues such as algorithmic bias, privacy concerns, and data security to build trust in AI and ML applications (Hansen et al., 2015).

Navigating the ethical landscape: A review of transparent and accountable AI systems in healthcare

The integration of Artificial Intelligence (AI) in healthcare promises significant advancements in patient care, diagnosis, and operational efficiency. However, concerns regarding the transparency and accountability of AI systems have emerged. Transparency is allowing AI systems to clearly explain their choices to build confidence and comprehension. Accountability methods guarantee compliance with legal and ethical criteria. This literature review explores existing research on designing transparent and accountable AI systems in healthcare. By identifying challenges, ethical considerations, and promising approaches, we aim to contribute to the discourse on AI ethics and governance in healthcare. Ultimately, prioritizing transparency and accountability can optimize the benefits of AI while ensuring ethical integrity and patient safety.

Applications of AI in healthcare: Enhancing transparency and accountability

The integration of AI into healthcare carries vast potential for breakthroughs, but ethical considerations such as accountability, transparency, and explainability are paramount. Studies "Unlocking the Black Box" delve into the decision-making processes of AI used in medical imaging, aiming to build trust and understanding among healthcare professionals, ultimately leading to better patient care. Similarly, transparency is crucial in AI-assisted drug discovery to ensure the validity and safety of new therapies. Additionally, AI-powered treatment suggestions and predictive analytics promise to revolutionize patient care, but clear and understandable AI processes are essential for healthcare professionals to grasp and trust these recommendations. By prioritizing ethical standards and accountability, these examples showcase responsible AI implementation in healthcare, maximizing benefits while safeguarding patient well-being (Hlávka, 2020).

In Table 4.1 we have presented different use cases with description.

Table 4.1 Use cases with description.

Use cases	Description
Clinical decision support systems	AI systems assist healthcare providers in making evidence-based decisions by providing transparent explanations for recommendations and predictions
Personalized treatment recommendation	AI can generate personalized treatment plans by analyzing patient data
Drug discovery and development	AI speeds up drug discovery by analyzing molecular structure and medical trial data
Medical imaging analysis	AI driven systems improve diagnostic accuracy in medical screening to validate diagnostics decisions
Healthcare resource allocation	Using AI driven predictive analytics helps make resource allocation more efficient
Patient engagement and education	Conversational AI interfaces keep patients engaged by offering transparent health advice and answering medical queries

Accountability, transparency and explainability in AI for healthcare

Megaro (2023) did research in the area that how AI into healthcare offers significant prospects for utilizing health data to enhance healthcare delivery, but it necessitates addressing governance challenges to ensure accountability. A literature review covering the years 1986–2021 revealed a gap in publications due to the "AI winter" from the late 1980s to the early 2010s. Notably, issues raised in papers from 1986, such as establishing accountability guidelines for clinical software and concerns about opaque algorithms, remain pertinent today. Recent policy initiatives aim to regulate AI use in high-risk areas, emphasizing the importance of mitigating unintended negative consequences. The review underscores a lack of clear definitions in the literature regarding accountability and the interplay between transparency, explainability, and accountability. While empirical research on AI governance is limited, future studies could assess proposed guidelines and frameworks. In essence, the review offers insights for researchers to contribute to advancing AI governance in healthcare.

Approximately 50% of the reviewed articles are focused on proposing solutions to ensure accountability for AI in healthcare (12 out of 21 articles). These solutions encompass design principles, technical components such as algorithms and software applications, and ethical guidelines aimed at addressing the challenges (Kiseleva, 2020).

Understanding explainability in cardiac arrest prediction tools

Amann (2022) research paper examines the consequences of including or excluding explainability in a cardiac arrest prediction tool utilized by dispatchers. It presents two socio-technical scenarios: one where explainability is absent and another where it is integrated. Scenario 1 delves into the hurdles associated with validating and developing a black-box system, considering technical aspects, human factors, and its role in decision-making. Scenario 2 outlines the potential advantages of incorporating explainability, along with corresponding technical and human considerations, and its impact on decision-making. The paper delves into the ongoing discourse surrounding the necessity of explainability in clinical decision support systems (CDSSs) and underscores the importance of tailored approaches tailored to specific use cases and user requirements. Additionally, it emphasizes the necessity of standardized terminology and interdisciplinary collaboration to address the complexities of explainability in AI-powered CDSSs. Ultimately, the conclusion suggests that the inclusion of explainability in CDSSs should be determined through individualized assessments, considering factors such as technical feasibility, validation processes, contextual nuances, user roles, and stakeholder viewpoints. The paper advocates for studying real-world applications to anticipate ethical and practical challenges and proposes an interdisciplinary approach to assess AI-powered CDSSs.

Unlocking the Black Box: exploring interpretability methods for deep neural networks in medical image analysis

Transparency of deep neural networks is an essential clinical, legal, and ethical requirement.The absence of understanding of deep neural networks' (DNNs') decision-making processes is a problem for integrating DNNs into clinical workflows for medical image processing tasks, as this review article discusses. The authors provide an overview of the several interpretability techniques that may be used to improve DNN transparency in medical image analysis, highlighting the significance of quantitative and qualitative assessments in order to provide reliable and solid justifications. They emphasize the need for application-grounded assessments including medical experts to gauge the effectiveness of interpretability techniques and identify any prejudices brought about by justifications that can result in an over- or underdiagnosis. The validation of interpretability techniques with imaging biomarkers and their possible contribution to the identification of novel biomarkers are also covered in the research. The road map for the deep

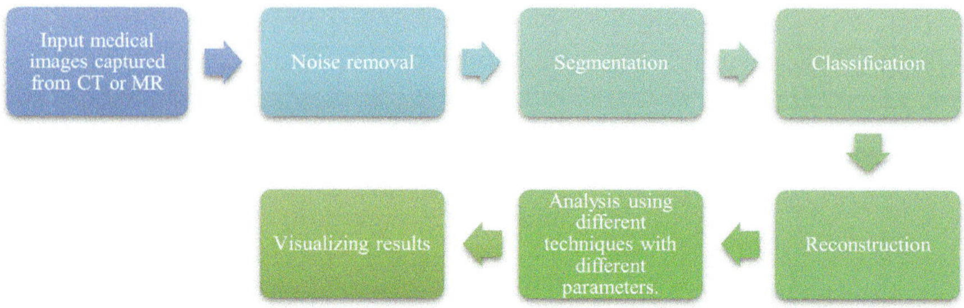

Figure 4.4 Roadmap of deep neural network roadmap of deep neural network.

neutral network is shown below in Fig. 4.4 showing all the steps involved from input the image to visualizing the results (Oh et al., 2021).

AI-aided drug discovery efforts

- AI in medicinal chemistry aids in predicting drug compound efficacy and toxicity, overcoming the limitations of traditional labor-intensive experimentation (Parashar et al., 2021).
- ML algorithms analyze vast datasets to identify patterns and trends, enabling faster identification of bioactive compounds with minimal side effects compared with classical protocols.
- Example: DL algorithm trained on known drug compounds accurately predicts the activity of novel compounds.
- ML algorithms also contribute to preventing potential drug compound toxicity by extensive training on databases containing toxic and nontoxic compounds.
- AI identifies drug-drug interactions by analyzing large datasets, leading to accurate prediction of interactions for novel drug pairs.
- Example: ML algorithm predicts interactions of novel drug pairs based on known drug interaction datasets.
- AI facilitates personalized medicine by identifying potential drug-drug interactions based on individual patient characteristics, enabling tailored treatment plans (Salahuddin et al., 2022).
- Overall, AI integration in pharmaceutical research improves prediction of drug efficacy and toxicity, leading to the development of safer and more effective medications and accelerating the drug discovery process.

Numerous case studies highlight the efficacy of AI in drug discovery. For instance, Gupta et al. successfully employed a Deep Learning (DL) algorithm to uncover novel cancer treatment compounds by training it on a comprehensive dataset of cancer-related compounds and their biological activity

(Santín et al., 2021). Similarly, ML facilitated the identification of small-molecule inhibitors targeting the MEK protein, a potential cancer treatment avenue. Additionally, AI algorithms have been instrumental in discovering inhibitors for beta-secretase (BACE1), implicated in Alzheimer's disease progression, and new antibiotics, including one effective against tuberculosis and resistant bacterial strains. In the context of COVID-19, ML has swiftly identified promising drug candidates from extensive compound datasets, expediting the drug discovery process significantly. These instances, alongside numerous others, underscore the capacity of AI-driven approaches to accelerate drug discovery and enhance medication efficacy.

Transforming healthcare: Leveraging artificial intelligence in clinical practice

Alowais et al. (2023) review article provides an up-to-date overview of AI's utilization within clinical practice, encompassing its potential roles in disease diagnosis, treatment suggestions, and patient interaction It also explores the challenges that come with using AI, such as moral and legal issues, highlighting the importance of human knowledge. The purpose of this article is to help healthcare organizations successfully integrate AI technology and to increase understanding of the role AI plays in healthcare. Fig. 4.5 depicts the evolution of AI in healthcare. It shows how AI has changed over time, starting with early rule-based systems and ending with the current era of machine learning and deep learning algorithms.

The promise for transforming healthcare through improved illness detection, laboratory testing, population health management, and drug optimization exists with the integration of artificial intelligence (AI) in clinical practice. Notwithstanding its potential, issues including public trust, cybersecurity, and regulatory norms must be resolved. Responsible AI adoption necessitates stakeholder collaboration and research and development expenditures. Artificial intelligence (AI) algorithms can help with preventative and treatment efforts by identifying high-risk individuals and predicting healthcare occurrences. All things considered, a smooth integration of AI is anticipated to transform the delivery of healthcare, improving patient outcomes and granting access to tailored treatment (Singh et al., 2023a).

The use of AI in healthcare: Past, present, and future

Hansen et al. (2015) presented the use of artificial intelligence (AI) in healthcare, stressing both the possible advantages and disadvantages. It highlights the growing need for ethical AI development in the healthcare

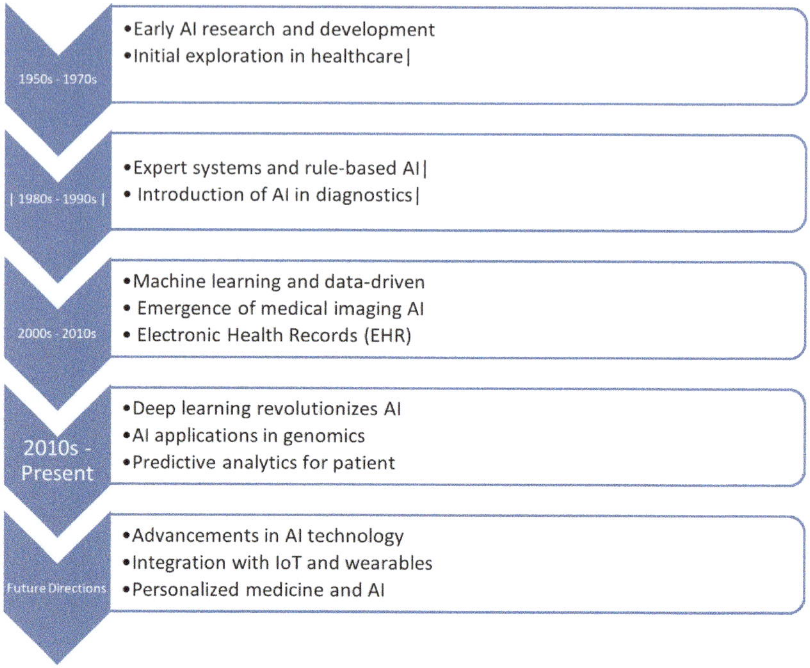

Figure 4.5 Evolution of AI in healthcare evolution of AI in healthcare.

industry and how crucial it is to match AI systems with moral principles in order to prevent any consequences. They present a workable paradigm for integrating ethics into AI for healthcare, which is based on a thorough analysis of current methodologies, ethical standards, and technological advancements. This framework emphasizes concrete solutions to empower developers and spans the whole AI lifecycle, from data management to model building, deployment, and monitoring.

Furthermore, they call for the creation of ethical AI checklists in collaboration with healthcare AI practitioners in order to fine-tune and execute the technological solutions offered. Finally, the study calls for a proactive approach to supporting human values and ethics in AI in healthcare in order to maximize advantages while limiting hazards.

Frameworks for designing transparent AI systems

In AI development within healthcare, there's a rising worry regarding the absence of transparency and accountability. To tackle this, several frameworks have been suggested, aiming to ensure the creation of transparent AI

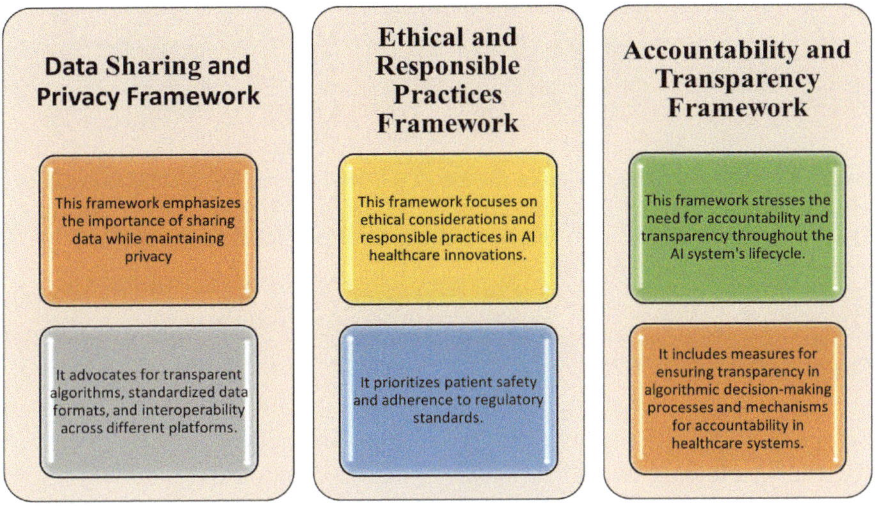

Figure 4.6 Different frameworks for the AI systems different frameworks for the AI systems.

systems in healthcare. One framework highlights the significance of data sharing, privacy, algorithm transparency, standardized data, and interoperability across various platforms. Another framework underscores the importance of ethical and responsible practices in AI healthcare advancements, prioritizing patient safety and regulatory compliance as shown below in Fig. 4.6 (Singh et al., 2023b).

Enhancing biomedical data sharing and privacy framework

Elkhodr et al. (2023) provides some insides about this digital healthcare era, sharing biomedical data is crucial for research and personalized healthcare, yet preserving privacy and security remains challenging. In their study, they offer BioChainReward (BCR), a system built on blockchain technology that addresses these issues. Better security, privacy, and incentives for exchanging biological data are guaranteed by BCR. Four layers make up its architecture: application, blockchain, data, and smart contracts. The blockchain layer controls hashing and retrieval, whereas the data layer encrypts and decrypts data. An AI-driven privacy-preserving component of the smart contract layer chooses the best methods for processing data requests and uses rewards to encourage data sharing. Lastly, the application layer provides access to shared data for various applications such as AI-enabled apps and data analysis tools. BCR offers a robust, comprehensive

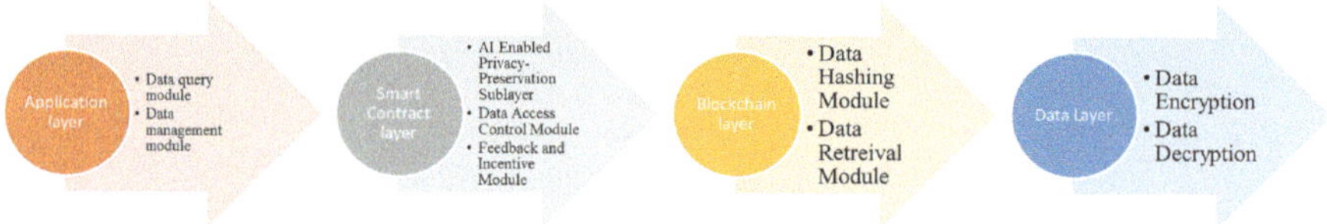

Figure 4.7 Framework architecture for the BCR. *(From Santín, E.P., Solana, R.R., García, M.G., Suárez, M.D.M.G, Díaz, G.D.B., Cabal, M.D.C., Rojas, J.M.M. & Sánchez, J.I.L. (n.d.). Toxicity prediction based on artificial intelligence: A multidisciplinary overview. Wiley Interdiscip. Rev. Comput. Mol. Sci, 2021.)*

solution for secure, privacy-aware, and incentivized biomedical data sharing (Singh and Singh, 2023).

The research paper presents findings from collaborative work between Santín, E.P. and the research group (GRP). It introduces BioChainReward (BCR), a new framework refer Fig. 4.7 utilizing blockchain technology to facilitate secure sharing of biomedical data with external entities such as research institutions. BCR comprises four layers leveraging blockchain advantages for data security during processing and transmission (Salahuddin et al., 2022).

BCR's notable feature is its integration of smart contract technology, incorporating an AI-driven privacy-preservation sublayer. This sublayer enables application of suitable privacy-preserving techniques based on specific data request purposes and nature. This customization not only ensures tailored security for each transaction but also grants users significant control over privacy settings, addressing major data security concerns.

Furthermore, BCR introduces a feedback incentive mechanism aiming to motivate and reward patients for sharing personal data. This approach seeks to enhance user engagement and participation while enriching the pool of available data for biomedical research.

Condensing and rephrasing the ethical and responsible practices framework

Guidelines for developers to implement ethical principles in AI systems for healthcare.

Creating a practical approach for incorporating ethical considerations into healthcare-focused artificial intelligence: guidance tailored for developers of AI systems (Singh et al., 2022). Solanki et al. (2023) have done deeper research and below are the findings for the approach and the principles to follow in AI in healthcare. Table 4.2 shows the AI ethics in healthcare industry.

Table 4.2 Navigating AI ethics in healthcare: A summary.

Domain	Description
Balancing potential and risks	AI's promise in healthcare must be weighed against potential risks
Ethical Imperative	There's a growing demand to ensure AI in healthcare is ethically aligned, addressing potential harms and respecting human values.
Practical ethical framework	The proposed framework offers actionable solutions for implementing ethics in AI for healthcare, bridging theoretical guidance with practicality
Informed by comprehensive review	Informed by a thorough review, the framework integrates existing ethical guidelines and technical approaches, emphasizing values such as self-direction.
Holistic lifecycle Coverage	Covering the entire AI lifecycle, from data management to model development and monitoring, ensures comprehensive ethical considerations.
Emphasis on Actionability	The focus lies on actionable solutions, both technical and nontechnical, to empower developers in integrating ethics into routine work.
Rooted in literature	Recommendations are grounded in existing literature, ensuring accessibility and adoption by AI developers.
Support for ethical checklists	Advocating for 'ethical AI checklists' codesigned with practitioners, facilitating the refinement and implementation of proposed technical solutions.

Accountability and transparency framework

Vourganas et al. (2022) presented the notion of transparency in the context of artificial intelligence (AI) in healthcare. They begin by looking at the idea of transparency as it is defined in EU legislation and AI policy documents, and then create a hierarchy of related notions such as explainability and interpretability. Transparency is regarded as the highest notion, with additional methods used to accomplish it.

Moving on to the healthcare setting, they investigates the roles and kinds of AI transparency, highlighting its importance as a system of accountability for all parties involved. This involves considering the risk-benefit approach, which is already popular in healthcare.

The examination extends to AI's outward transparency toward patients, concentrating on the legal framework of medical informed consent. It then delves into the internal and insider transparency of AI towards healthcare providers and developers, respectively, with attention to the EU's Medical Devices Regulation (MDF).

They provide the current literature by addressing the rules and regulations that govern AI development and usage in healthcare, bridging gaps between medical informed consent and the MDF, and identifying links between transparency requirements. It presents transparency as a tiered structure of accountability and highlights weaknesses within it, proposing solutions.

This research thoroughly analyzes how transparency is approached in the integration of AI within healthcare, covering its significance, functions, legal considerations, and avenues for improvement. It helps further our understanding of how transparency can be effectively implemented in AI applications within the healthcare sector.

Conclusion

The integration of artificial intelligence (AI) with healthcare represents a dramatic prospect for improving patient care, diagnostics, and operational efficiency. However, questions about transparency and accountability in AI systems have surfaced as key issues. This paper investigated the significance of developing transparent and accountable AI systems in healthcare by examining the present level of AI adoption, advantages, obstacles, and techniques for fostering openness and accountability.

This study provides significant insights for healthcare businesses trying to manage AI deployment while emphasizing patient safety, privacy, and trust. The study emphasizes the importance of standards, tactics, and ethical concerns in establishing transparent and responsible AI systems for healthcare.

Through a thorough examination of case studies, ethical frameworks, regulatory guidelines, and existing literature, this research provides valuable insights for healthcare organizations seeking to navigate AI implementation while prioritizing patient safety, privacy, and trust. The paper underscores the need for standards, strategies, and ethical considerations in developing transparent and responsible AI systems in healthcare.

Furthermore, various use cases, such as AI-assisted medical imaging analysis and drug discovery efforts, exemplify the importance of transparent AI processes in building trust among healthcare professionals and ensuring the effectiveness and safety of AI-driven interventions. By upholding ethical standards and accountability, responsible AI implementation in healthcare can maximize benefits while safeguarding patient well-being.

In response to the challenges posed by the lack of transparency and accountability in AI development, several frameworks have been proposed. These frameworks emphasize data sharing, privacy, algorithm transparency, and ethical practices to ensure the creation of transparent AI systems in healthcare (Alowais et al., 2023; Amann, 2022; Drysdale, 2019; Elkhodr et al., 2023; Ethics and governance of artificial intelligence for health: WHO guidance, 2021; Felzmann et al., 2020; Gupta et al., 2021; Hansen et al., 2015; Hlávka, 2020; Hond and Anne, 2022; Kiseleva, 2020; Megaro, 2023; Oh et al., 2021; Parashar et al., 2021; Salahuddin et al., 2022; Santín et al., 2021; Shick et al., 2024; Singh and Kant Singh, 2024; Singh et al., 2023a,b, 2022; Singh and Singh, 2023; Solanki et al., 2023; Sun, 2023; Tekkesin, 2019; Venkatesh et al., 2022; Vourganas et al., 2022). Moving forward, continued efforts to prioritize transparency, accountability, and ethical integrity are essential for harnessing the full potential of AI in healthcare while safeguarding patient welfare.

References

Alowais, S.A., Alghamdi, S.S., Alsuhebany, N., Alqahtani, T., Alshaya, A.I., Almohareb, S.N., Aldairem, A., Alrashed, M., Bin Saleh, K., Badreldin, H.A., Al Yami, M.S., Al Harbi, S., Albekairy, A.M., 2023. Revolutionizing healthcare: the role of artificial intelligence in clinical practice. BMC Med. Educ. 23 (1). https://doi.org/10.1186/s12909-023-04698-z.

Amann, J., 2022. To explain or not to explain?—artificial intelligence explainability in clinical decision support systems. PLOS Digit. Health 1.

Drysdale, E., 2019. Vector-SickKids Health AI Deployment Symposium.

Elkhodr, M., Gide, E., Darwish, O., Al-Eidi, S., 2023. BioChainReward: a secure and incentivised blockchain framework for biomedical data sharing. Int. J. Environ. Res. Publ. Health 20 (19), 6825. https://doi.org/10.3390/ijerph20196825.

Ethics and Governance of Artificial Intelligence for Health: WHO Guidance, 2021.

Felzmann, H., Fosch-Villaronga, E., Lutz, C., Tamò-Larrieux, A., 2020. Towards transparency by design for artificial intelligence. Sci. Eng. Ethics 26 (6), 3333–3361. https://doi.org/10.1007/s11948-020-00276-4.

Gupta, R., Srivastava, D., Sahu, M., Tiwari, S., Ambasta, R.K., Kumar, P., 2021. Artificial intelligence to deep learning: machine intelligence approach for drug discovery. Mol. Divers. 25 (3), 1315–1360. https://doi.org/10.1007/s11030-021-10217-3.

Hansen, K., Biegler, F., Ramakrishnan, R., Pronobis, W., Von Lilienfeld, O.A., Müller, K.R., Tkatchenko, A., 2015. Machine learning predictions of molecular properties: accurate many-body potentials and nonlocality in chemical space. J. Phys. Chem. Lett. 6 (12), 2326–2331. https://doi.org/10.1021/acs.jpclett.5b00831.

Hlávka, J.P., 2020. Security, privacy, and information-sharing aspects of healthcare artificial intelligence. In: Artificial Intelligence in Healthcare, pp. 235–270. https://doi.org/10.1016/B978-0-12-818438-7.00010-1.

Hond, de, Anne, A.H., 2022. Guidelines and quality criteria for artificial intelligence-based prediction models in healthcare: a scoping review. NPJ Digit. Med. 5.

Kiseleva, A., 2020. AI as a medical device: is it enough to ensure performance transparency and accountability? Eur. Pharmaceut. Law Rev. 4 (1), 5–16. https://doi.org/10.21552/eplr/2020/1/4.

Megaro, A., 2023. Transparency in AI systems for value co-creation in healthcare. In: Big Data and Decision-Making: Applications and Uses in the Public and Private Sector, pp. 93–105. https://doi.org/10.1108/978-1-80382-551-920231007.

Oh, J.S., Kim, M.K., Lee, D.H., 2021. A study on the selection of future AI+X promising fields and the direction to strengthen competitiveness. In: 3rd International Conference on Artificial Intelligence in Information and Communication, ICAIIC 2021, pp. 371–374. https://doi.org/10.1109/ICAIIC51459.2021.9415181.

Parashar, G., Chaudhary, A., Rana, A., 2021. Systematic mapping study of AI/machine learning in healthcare and future directions. SN Comput. Sci. 2 (6). https://doi.org/10.1007/s42979-021-00848-6.

Salahuddin, Z., Woodruff, H.C., Chatterjee, A., Lambin, P., 2022. Transparency of deep neural networks for medical image analysis: a review of interpretability methods. Comput. Biol. Med. 140. https://doi.org/10.1016/j.compbiomed.2021.105111.

Santín, E.P., Solana, R.R., García, M.G., Suárez, M.D.M.G., Díaz, G.D.B., Cabal, M.D.C., Rojas, J.M.M., Sánchez, J.I.L., 2021. Toxicity prediction based on artificial intelligence: a multidisciplinary overview. Wiley Interdiscip. Rev. Comput. Mol. Sci. 11 (5).

Shick, A.A., Webber, C.M., Kiarashi, N., Weinberg, J.P., Deoras, A., Petrick, N., Saha, A., Diamond, M.C., 2024. Transparency of artificial intelligence/machine learning-enabled medical devices. Npj Digit. Med. 7 (1). https://doi.org/10.1038/s41746-023-00992-8.

Singh, A., Dhull, A., Singh, K.K., Singh, K.K., 2023a. AIML-Based blockchain solutions for IoMT. In: Blockchain and Deep Learning for Smart Healthcare, pp. 73–93. https://doi.org/10.1002/9781119792406.ch4.

Singh, A., Dhull, A., Singh, K.K., Singh, K.K., 2023b. Blockchain in healthcare. In: Blockchain and Deep Learning for Smart Healthcare, pp. 147–169. https://doi.org/10.1002/9781119792406.ch7.

Singh, A., Singh, K.K., Greguš, M., Izonin, I., 2022. CNGOD-An improved convolution neural network with grasshopper optimization for detection of COVID-19. Math. Biosci. Eng. 19 (12), 12518–12531. https://doi.org/10.3934/mbe.2022584.

Singh, A., Kant Singh, K., 2024. Blockchain and deep learning: research challenges, open problems, and future. In: Blockchain and Deep Learning for Smart Healthcare, pp. 425–431.

Singh, A., Singh, K.K., 2023. FedDDR: A Federated Improved DenseNet for Classification of Diabetic Retinopathy, 2023.

Solanki, P., Grundy, J., Hussain, W., 2023. Operationalising ethics in artificial intelligence for healthcare: a framework for AI developers. AI Ethics 3 (1), 223–240. https://doi.org/10.1007/s43681-022-00195-z.

Sun, L., 2023. Artificial intelligence for healthcare and medical education: a systematic review. Am. J. Tourism Res. 15.

Tekkesin, A.İ., 2019. Artificial intelligence in healthcare: past, present and future. Anatol. J. Cardiol. 22. https://doi.org/10.14744/anatoljcardiol.2019.28661.

Venkatesh, C., Rengarajan, N., Ponmurugan, P., Balamurugan, S., 2022. AI in healthcare. In: Smart Systems for Industrial Applications, pp. 115–140. https://doi.org/10.1002/9781119762010.ch5.

Vourganas, I., Attar, H., Michala, A.L., 2022. Accountable, Responsible, Transparent Artificial Intelligence in Ambient Intelligence Systems for Healthcare. Springer Nature.

CHAPTER 5

Ensuring fairness and mitigating bias in healthcare AI systems

Akansha Singh[1] and Krishna Kant Singh[2]
[1]School of CSET, Bennett University, Greater Noida, Uttar Pradesh, India; [2]Delhi Technical Campus, Greater Noida, Uttar Pradesh, India

Introduction to fairness in healthcare AI

Artificial Intelligence (AI) is rapidly revolutionizing the field of healthcare, in terms of both its diagnostic and treatment personalization capacities, and its potential for efficient operation. With global health systems increasingly adopting AI technologies, fairness in the implementation of AI-based solutions within this sector has become a high-profile concern. In the context of healthcare, fairness in AI encompasses the ethical responsibility to prevent, identify and mitigate biases that may be introduced at various stages of the development and deployment of AI systems. Biases may arise from one or more of a range of sources, including imbalances in the training data, the design of algorithms, or the misinterpretation by the healthcare players of AI outputs.

Moreover, the fairness commitment goes farther than just technical adjustments—it is complex, holistic, and covers ethical considerations and regulatory compliance, involving the proactivity of very many stakeholder groups to take proper care that the AI systems do not aggravate health disparities already existing or create new ones. For instance, an AI diagnostic system developed on primary data from one ethnic group is likely to have low performance when applied to another ethnic group, probably leading to poor diagnoses or underdiagnoses. In a similar manner, socioeconomic biases may lead to AI healthcare solutions being less accessible to the most underprivileged, who often need innovative healthcare solutions the most.

Solutions to make healthcare AI fair, therefore, need to include multidimensional strategies, comprised of diversifying the training data, designing inclusive algorithms, and ensuring that such work is anchored

Responsible and Explainable Artificial Intelligence in Healthcare
ISBN 978-0-443-24788-0
https://doi.org/10.1016/B978-0-443-24788-0.00005-4

within robust governance frameworks through which its transparency and accountability may be guaranteed. This has to be done with the involvement of ethicists, community representatives, and interdisciplinary experts in guiding the development and review processes so that such AI technologies are more likely to align with broader social values and goals for health care. Ultimately, a concern with fairness in healthcare AI is about much more than upholding the fundamental medical ethic of "do no harm." It is about seeing that the technology is fair to every population. This commitment to equity will help in building the much-needed trust of AI applications, developing wider acceptance, and truly harnessing the potential of AI to improve health outcomes globally.

The role of AI in healthcare

Artificial intelligence is entering different aspects of health care, from predictive diagnostics to personalized medicine. Run by complex algorithms and machine learning, AI systems can handle huge loads of patient data, imaging, or genetic information in decision-making pertinent to clinical practice. Artificial Intelligence has significantly impacted the healthcare industry, transforming the conventional practices into more innovative and effective methodologies. This reflection is made in several key areas.

- **Predictive Diagnostics:** The contribution of AI for scanning through and analyzing the big data samples in the field of diagnostics cannot be overlooked. Machine learning models have the ability to find patterns and anomalies that could be indicative of diseases with more accuracy than humans (Esteva et al., 2019). Specifically, such predictive capabilities have proven to be quite effective at detecting conditions early, such as in the case of cancer, where AI algorithms scan imaging data to detect malignancies at stages where they can most easily be defeated.
- **Personalized Medicine:** Personalized medicine stands to gain a lot from AI in tailoring healthcare to individual patient characteristics. With the help of such genetic information, the technology assists in the prediction of how an individual will respond to various treatments and hence a more adjusted form of caring (Joyce and Topol, 2019). That side of AI in healthcare promises to bring better outcomes by trial and error reduction in medicine prescription.
- **Clinical Decision Making:** Clinical decision-making is gradually becoming an evidence-based process due to the adaptation of AI. These

algorithms digest extensive patient data in making well-informed decisions, right from electronic health records to clinical notes, for the provision of support for health care providers (Ghassemi et al., 2018). This data-driven approach has made it easy and has played a huge role in refining options for treatment and strategies for new therapeutics.

- **Cost-Efficiency and Reduction:** Applications of AI make administrative tasks in health services, such as scheduling, billing, and patient flow management, more efficient, cutting down on operational costs. This eliminates or lessens the value of routine automation and allows the health workforce to work on more demanding aspects of care.
- **Remote Monitoring and Telemedicine:** In this regard, AI has been lying at the core of this recent surge of telemedicine, enabling healthcare providers to monitor the health statuses of patients. With the use of wearable devices and mobile applications that have been incorporated with AI, health data is analyzed, and if there are possible health threats, alerts are sent to healthcare providers before the patients could complain of symptoms.

The promise of AI: Enhancements in patient care

The application of AI in health settings, and particularly in healthcare, has been so promising in bettering patient care through better and more accurate diagnoses, optimal treatment, and prediction of patient outcomes with new levels of precision that were not possible before. AI could offer the best chances for early detection in diseases such as cancer, often identifying subtle patterns in data that might escape human eyes. Furthermore, AI applications in the coordination of patient care logistics can bring about more efficient ways of delivery and better patient experiences. It is the incorporation of artificial intelligence in healthcare that is seen as the beacon of progress, an indication that soon there will be the use of more nuanced and sophisticated modes of patient care. Thus, AI holds a lot of promise with many, if not most, of the fundamental features reflecting the new wave of a great leap forward in healthcare.

- **Improved Diagnostic Accuracy:** The first promise of AI is in its computational power to augment diagnostic accuracy. Algorithms trained on large and diverse datasets can decode intricate diagnostic markers, thereby enabling early and accurate detection of diseases. It is most relevant in conditions where early interventions significantly improve prognosis, such as in most cancers.

- **Personalized Treatment Optimization:** Personalized treatment is yet another main benefit of AI in health. In other words, AI can help in developing protocols of treatment through data crunching on individual patients with a view to support personalized care. It takes into consideration the genetic makeup and lifestyle of the patient in order to optimize the therapeutics for efficiency. The personalized approach to medical treatment is a radical deviation from the medical treatment strategy, "one size fits all," which has prevailed in medicine.
- **Prognostic Precision:** AI prognosticates at a very strong level, hence being able to predict patients' outcomes with precision. This gives caregivers the ability to make decisions informed by a rational long-term care strategy, yet at the same time, frame a realistic view with patients and their families.
- **Healthcare Logistics Enhancement:** AI is redefining logistics in the entire patient care sector. In this manner, the function of healthcare systems is being made efficient by the reduction in administrative burdens, the management of patient flow, and resource allocation to the best optimization. In simpler terms, the said improvement in health care logistics helps to save time and resources and even enhances the patient experience as a whole due to a reduction in the waiting time and increased accessibility of care services.
- **Enabling Remote Care:** The use of AI is most prevalent in the area of remote patient monitoring and telehealth. The contribution of AI has been immense in ensuring remote monitoring of patient health to ensure that medical care continues unabated without risking the spread of the infection during the period of the COVID-19 pandemic. This has established the role AI will have in delivering care outside the traditional clinical environment (Fig. 5.1).

The challenge of bias: Risks and consequences

However, the deployment of AI in healthcare is not without significant challenges. The existence of bias in AI systems poses substantial risks that can lead to inequitable treatment and outcomes for patients (Gianfrancesco et al., 2018). Biased AI can arise from nonrepresentative training data, leading to less accurate predictions for underrepresented groups, which in turn can perpetuate and amplify existing healthcare disparities (Rajkomar et al., 2018). Despite the groundbreaking potential of Artificial Intelligence (AI) in healthcare, the specter of bias casts a long shadow over its

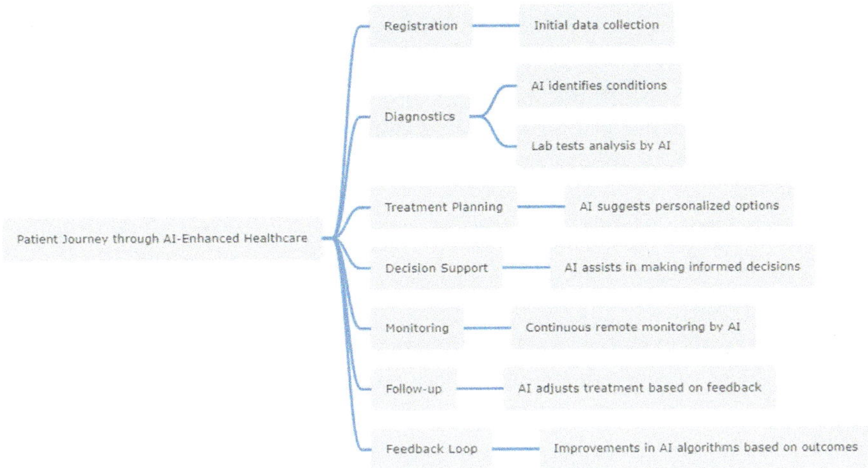

Figure 5.1 Patient journey through AI enhanced healthcare.

deployment. The risks and consequences associated with biased AI systems are multifold and can significantly undermine the goal of equitable healthcare. The genesis of bias in AI often lies in the data it is fed. AI systems learn to make decisions based on the data they are trained on, and if this training data is not representative of the diverse patient population, the AI's predictions may be skewed towards the majority group represented in the dataset (Obermeyer et al., 2019). This situation creates a feedback loop where historical inequalities are not only perpetuated but potentially worsened. When bias infiltrates AI systems in healthcare, it can manifest as inequitable treatment and outcomes. For instance, a study revealed that an algorithm used in the US health system was less likely to refer Black people than white people who were equally sick to programs that aim to improve care for patients with complex medical needs (Obermeyer et al., 2019). This reflects the risk of exacerbating disparities in healthcare access and quality. Bias in AI has the potential to compromise the integrity of public health initiatives. Inaccurate predictions for certain groups can lead to poor health outcomes, higher rates of misdiagnosis, and a loss of trust in healthcare systems, which is particularly damaging in communities that already face healthcare inequities (Vyas et al., 2020). There are also ethical and legal implications to consider. Healthcare providers have a moral obligation to deliver care that is fair and just. The use of biased AI tools conflicts with these ethical standards and could lead to legal ramifications if they result in discriminatory practices (Char et al., 2020). To address these challenges, it is crucial to implement strategies for identifying and mitigating bias in AI systems. This includes diversifying datasets, implementing algorithmic fairness techniques, and ensuring continuous monitoring and evaluation of AI systems in clinical settings (Chen et al., 2018). Top of Form

The importance of fair AI systems

It cannot be overstressed enough that there is urgency toward fairness in AI so that, in this case, the technology is harnessed to its fullest capability within the health system. AI must be fair to ensure fair access to healthcare benefits for the highly diverse populations differentiated by races, genders, and socio-economic statuses all over the world. The importance of fairness within healthcare AI cannot be understated, being the basis on which AI can be considered to be applied ethically in a domain as sensitive as healthcare, where the stakes belong to human beings and their well-being. The importance of deploying fairness within AI systems for healthcare cannot be overstated. In a fundamentally human-centric domain such as healthcare, the consequences of employing AI systems transcend merely operational efficiency; they reach to the very ethical fabric of medical practice and patient welfare.

Fairness is not a technical goal of healthcare AI but an ethical imperative. AI systems in health must be made to accord with the principles of beneficence, nonmaleficence, autonomy, and justice that guide the rest of medical ethics. In this way, the benefits in terms of healthcare for every patient, irrespective of their background, are assured within a set of previously mentioned timeless principles in the era of digital medicine.

Fair AI systems are important to help achieve health service delivery equity. Such systems would be in a position to note and bridge the gaps in health service delivery, especially to those who are underserved. By ensuring fair insight and support, AI can make the service delivery care-attuned to the specific needs and circumstances of each group of patients, serving to advance more equitable health outcomes.

Trust is one of the cornerstones of healthcare, and fair AI systems are important for establishing trust between patients and technology. If patients are aware that they are being treated fairly by AI systems, this gives them more interest and better benefits in all AI-driven healthcare interventions. Trust is important for effective usage of AI technologies in healthcare.

The impact of fair AI systems is across individual care of the patient; it extends to societal well-being. Fair AI systems could be a potential contributor to social health equity, which is a determinant of social stability and cohesion in society at large. Fair AI systems, by not perpetuating disparities, could be a benefit to the broader social good.

The FAIR AI System also ensures the compliance of evolving regulations that govern the use of AI in healthcare. Regulators are also turning to ethics in AI deployment, and any noncompliance exposes to legal risks, as well as sanctions and reputation loss.

In an economic view, fair AI systems would allocate healthcare resources efficiently by reducing biases, which could lead to either unnecessary or ineffective treatments. AI channelizes resources where really needed, generally enhancing the cost-effectiveness of delivering healthcare.

Understanding bias in AI

Bias in the field of Artificial Intelligence can be described as the "systematic deviations in an algorithm's output that create unfair or unethical outcomes to one or more individuals or groups." The bias in AI arises from systematic prejudgment by an algorithm, based on false premises of the learning process. For example, it can be a result of information imbalance in the training data used for machine learning (Fig. 5.2).

Types of bias in AI systems

Most artificial intelligence bias can be generalized into three categories.

- Preexisting bias: This is when AI algorithms are trained on data that represents prior or existing societal inequalities. These are then fed into real-world data that are accumulated from human decisions or societal structures and, as a result, are learned by the AI.
- Technical Bias: Technical bias arises from the working mechanism of AI systems. These include inadequacies within the algorithm itself, such as model oversimplification, or within the mechanisms around sampling and preparing data, which cause a misrepresentation of the real world in the training data.
- Emergent Bias: This kind of bias is acquired over time by AI systems when they are exposed to new data or when applied to new or changing contexts. The problem is due to the relationship of the AI system with its users, and its outcomes could not have been predicted from the very first time of design and deployment of the system.

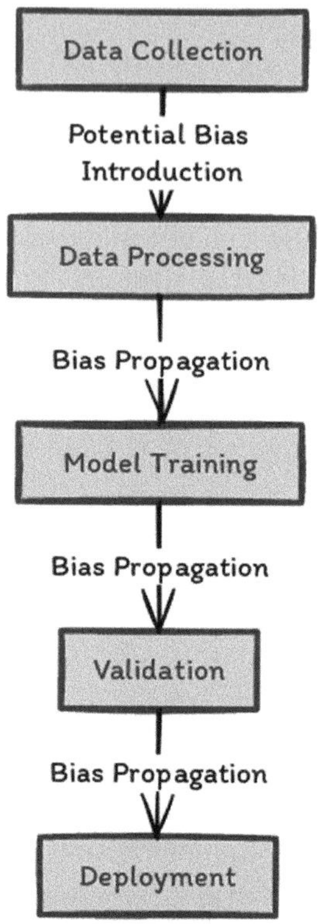

Figure 5.2 AI development life cycle.

Bias in AI systems is critically related to patient outcomes, to the extent that it may lead to misdiagnoses, recommendations for incorrect treatments, and unequal share in the access to care. At worst, these biases could actually serve to reinforce systemic health inequities that impact entire groups of populations, not only individual patients.

Sources of bias in healthcare AI systems

Bias in healthcare AI can certainly originate from a wide range of sources, and each contributes to possible disparities in performance of AI systems across populations. These are discussed in more detail below.

- **Data Bias:** If the training set for data modeling has no diversity or is formulated in such a way that it actually represents the entire population, the outcome is a biased model. For example, a data set with the majority of medical images coming from one ethnic group will most likely make the AI model underperform in disease diagnostics for people from other ethnic groups. These kinds of disparities can eventually lead to AI models inadvertently perpetuating and even widening health disparities.

- **Algorithmic Bias:** Bias inherently takes place within the mechanics of an algorithm. Even when diverse data are fed to algorithms, processing of the data may elicit bias in the outcomes. For example, one would ideally expect the classifier to have unbiased results since it has been optimized on a global error measure, but a classifier may well perform for the majority and do poorly for the minority at the subgroup level. Such biases are quite inadvertent and can arise out of factors such as the choice of model, the criteria of optimization, or in-built assumptions within the algorithm.
- **Label Bias:** The bias would be introduced in the data labeling process. These people may have some kind of conscious or unconscious bias, or probably cultural and social contexts have decided what is normal or abnormal, and hence their labeling would carry these biases in them. For example, there may be varying cultural understandings of pain, which therefore influence how symptoms get documented in a manner that will subsequently affect training data for diagnostic algorithms. This kind of bias is often hard to mitigate because that requires a deep understanding of the context in which the data is collected and labeled.

Addressing these biases is crucial to developing effective AI systems in healthcare that are fair and equal across all demographic groups. This will require not only technical solutions in the form of better algorithmic solutions and more inclusive data collection but also broader institutional changes in curating datasets and monitoring and regulating AI deployment across healthcare settings.

Identifying sources of bias

An understanding and identification of the sources of bias in healthcare AI are key to devising strategies to mitigate their impact on fairness. Bias in AI can stem from various stages of the AI development and deployment lifecycle. The following are the primary sources of bias that can affect the results of a healthcare AI system.

A. Data Collection and Bias

- **Representativeness Bias**: The model will, in most cases, not perform well with new data if it is not representative of the population. This kind of bias comes around if there is underrepresentation in the demographic groups within the training dataset. For instance, if a dataset was highly male-dominated, that would result in an AI system that is awful at predicting diseases in females.

- **Sampling Bias**: This occurs when the process for data collection is sampling from a population not representative of the diversity that exists within the target population. This will make an AI model run perfectly with sample data but misfire in applications based on the real world if the characteristics of the populations are different from those of the sample.
- **Historical Bias**: This is an embedded historical data by either past discriminating practices or societal norms. AI models are trained on such biases when trained on historical data. For instance, it may so happen that the existence of differences in treatment history along racial lines may impel an AI system to perpetuate these disparities.

B. Label Bias and Data Annotation

- **Subjective Interpretation**: Labels based on subjective judgment can very often create inconsistencies because of personal bias of the annotator or based on cultural background. For example, different radiologists are likely to give different interpretations to medical images because of differences in training and experience, thus creating bias in the dataset.
- **Error Bias**: Human error might get introduced to data by incorrect labeling. Such errors can guide the learning of the AI model in the wrong direction and thereby make conclusions about features or patterns that do not exist.

C. Training the Model and Algorithmic Bias

- **Algorithm Design**: In some other cases, algorithms may be designed to inherently have bias due to their mathematical structure or the process of optimization utilized in their design. For example, some might overfit to the majority class in an imbalanced data set, leading to poor performance on minority classes.
- **Feature Selection Bias**: The chosen features to train the AI model may have bias in them; therefore, they do not capture characteristics that would be significant in making a just prediction between different groups. Leaving out relevant features or including irrelevant ones could well lead to biased results.

D. Deployment and Feedback Loops

- **Feedback Loop Bias**: AI systems learn from new data through continuous learning in the deployment phase. If the new data contains certain biases, then the AI system could reinforce that bias over time, especially when the learning algorithms don't account for it.
- **Contextual Bias**: As changes are effected in the operational environment or in the characteristics of the population, that which was trained with could likely have not been present; hence, the possibility of this leading to biases. An AI system trained in one geographical location might be biased when deployed in another location with different demographic distributions.

Identifying those sources of bias would be the first step toward working to mitigate their effects in healthcare AI systems. Identifying the source of the bias allows developers and implementers to put in place remedies at the point where most biases are likely to originate, with targeted action in the quest for more equitable and efficient health solutions (Fig. 5.3).

Impact of bias

There are profound impacts of bias on AI-infused healthcare systems such as.

- **Inequitable Outcomes:** Bias can bring about inequity among different groups of patients stemming from a data imbalance or imperfection within the algorithm training process. This would thereby bring about less accurate diagnoses, less effective forms of treatment, or preventive measures relative to other demographic groups. For example, an AI system may be skewed in its training data towards members of one ethnic group and therefore provide poor outcomes for members of other ethnicities.
- **Erosion of Trust:** Effective implementation of AI in healthcare is basically going to pivot around the trust that patients and healthcare providers shall repose in them. The presence of any kind of biases—be it actual or perceived—can erode this trust vastly. The patients might,

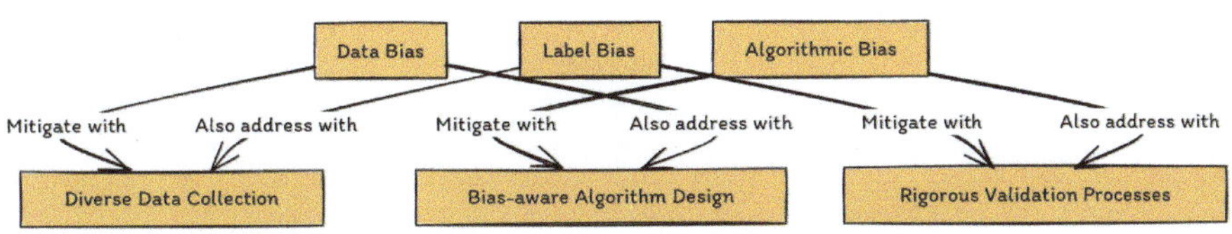

Figure 5.3 Bias sources and mitigation strategies.

therefore, be apprehensive about the fairness and reliability of care driven by AI, while healthcare professionals might be reluctant to rely on tools that would, in some situations, be linked with potential malpractice and ethical breaches. This erosion of trust will therefore work to prevent the adoption of potentially useful AI technologies and thus stall developments in the endeavor to improve healthcare efficiency and outcomes through innovation. The impact of bias in AI systems in healthcare can be profound (Fig. 5.4):

Strategies for mitigating bias

- **Collection of Diverse and Representative Data**:
  - **Broad Data Sources**: Ensure that the data used in training AI systems encompass wide demographics in terms of age, ethnicity, gender, medical history, among others. This will ensure the development of algorithms that can operate across patient populations.
  - **Continuous Data Monitoring**: Data sets reflecting changes in population dynamics and medical practices should be updated and monitored regularly to ensure relevance and fairness in AI applications.
- **Bias Detection and Assessment Techniques**:
  - **Algorithmic Audits**: Periodic audits of the AI algorithms may identify and correct any biases injected at the model-development stage. The objectivity of such assessments may be further improved by engaging independent third-party auditors.
  - **Transparent Reporting**: Encourage transparency in the reporting on how the data sets are constructed and how algorithms operate, including the limitations of both. This might be quite valuable in enabling peer review and broader forms of scrutiny to find biases that could otherwise be overlooked.

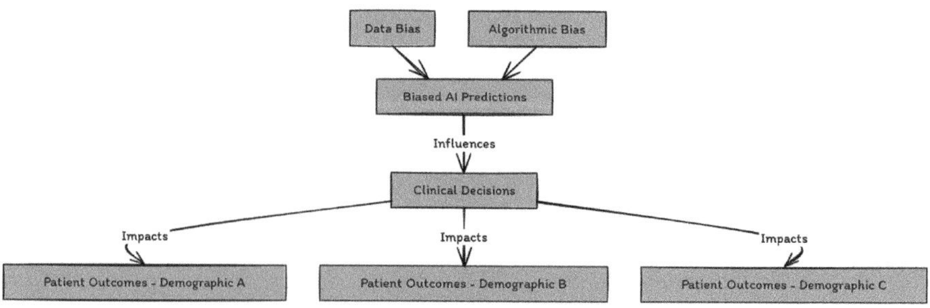

Figure 5.4 Impact of bias in AI systems in healthcare.

- **Advanced Algorithmic Design**:
  - **Debiasing Techniques**: Apply machine learning techniques that are specifically designed to automatically mitigate bias, such as reweighing training data, adversarial training methods, or fairness constraints.
  - **Multidisciplinary Approaches**: Ensure multidisciplinary approaches that have in place a wide range of perspectives and ethical considerations through the incorporation of different domain experts, among them, the experts in ethics, social science, and clinical medicine, during the design of the algorithm.
- **Regulatory and Ethical Frameworks**:
  - **Standards and Guidelines**: Developing and adhering to guidelines within the industry that are widely accepted and ethical guidelines, which explicitly deal with bias. This can be included as a protocol for all processes in AI development, validation, and deployment.
  - **Legal Compliance**: Ensure that all AI systems operate within the law and regulations of a country, such as the USA's HIPAA, which covers the domain of data privacy and security and comprises nondiscrimination laws.
- **Stakeholder Education and Involvement**:
  - **Training of Healthcare Providers**: Healthcare providers should be educated on the functionalities as well as the limitations of the AI tools. The training should also emphasize the required human supervision of decisions made by the AI.
  - **Patient Engagement**: Engage the patients in discussions on how the AI system will be used in providing care and the possible bias that could be evident in such use. This earns trust from the patient's side and also makes it possible for them to be included in treatment decisions.
- **Continuous Improvement and Adaptation**:
  - **Feedback Loops**: Create channels for receiving feedback from patients and health care providers regarding the performances of the applications of AI and make iterative improvements in the AI systems.
  - **Outcome Tracking**: Systematically track health outcomes from AI applications to ensure AI applications are improving patient care equitably among all demographic groups.

These strategies can be integrated by healthcare providers and AI developers to reduce bias in AI systems, thus improving fairness and efficacy in healthcare delivery and maintaining public trust in emerging technologies (Fig. 5.5).

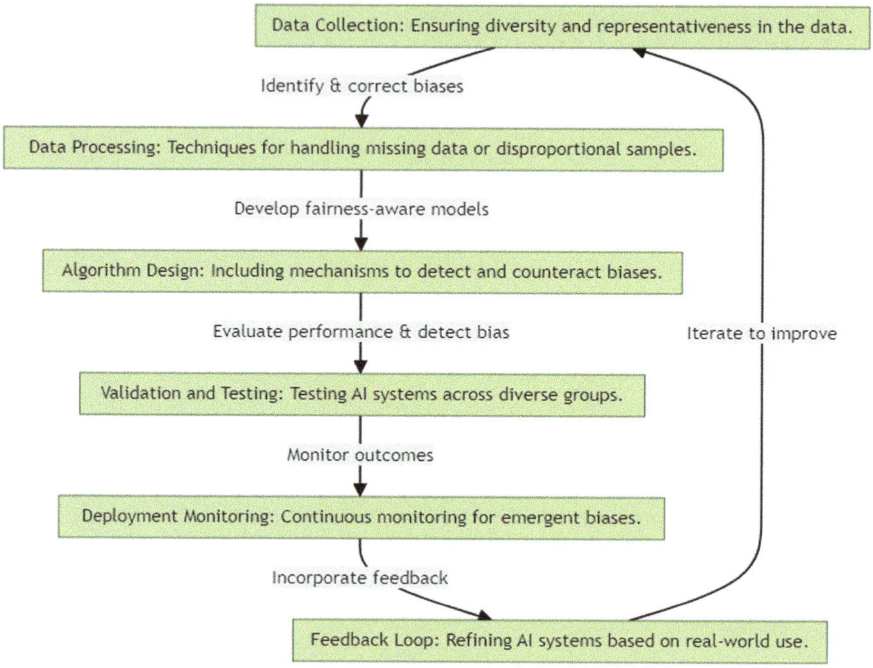

Figure 5.5 AI bias mitigation workflow.

Fairness in AI

Fairness in artificial intelligence is a multidimensional concept dealing with impartiality and equity in automation decision-making processes. In the context of AI, fairness is usually framed as no discrimination or favoring of an individual or group based on intrinsic or extrinsic characteristics such as race, gender, ethnic background, or socioeconomic status. AI systems must be designed to operate without discrimination, ensuring their decisions do not result in unjust or prejudiced outcomes.

The concept of fairness in AI can be dissected into several key dimensions.

- **Procedural Fairness:** Also known as process fairness, this relates to the methods and procedures used during the development and deployment of AI systems. It ensures that these processes are transparent, consistent, and accountable to all groups. Procedural fairness focuses on how decisions are made and whether the decision-making processes equally favor all individuals, irrespective of their group affiliations.

- **Distributive Fairness:** This dimension concerns the outcomes of AI decisions. It examines whether the benefits and burdens of AI technologies are distributed fairly among different groups. This aspect is critical in situations where AI decisions significantly impact individuals' lives, such as in healthcare, hiring, and law enforcement.
- **Interactional Fairness:** This form of fairness addresses the nature of the interactions between AI systems and their users. It emphasizes respect, politeness, and dignity in these interactions, ensuring that AI systems operate in ways that are perceived as fair and respectful by all users.
- **Representational Fairness:** This ensures all groups are equally represented in the training data and thus in the decision-making process of AI systems. It aims to prevent biases that might arise from underrepresentation or misrepresentation in the data from which AI learns.

Ensuring fairness in healthcare AI is crucial not only for ethical reasons but also for enhancing the effectiveness of healthcare delivery across all demographics. Achieving this requires understanding the sources of bias and implementing robust strategies to mitigate these biases. This endeavor demands a concerted effort among scientists, practitioners, and policymakers to uphold the principles of equity and justice in the rapidly evolving domain of healthcare technology.

Measuring fairness in AI systems

The measurement of fairness in AI involves various metrics and methodologies that quantitatively assess the fairness of an AI system across several dimensions.

- **Equality of Opportunity**: This metric ensures that all individuals have the same opportunities to achieve favorable outcomes from the AI system, regardless of their group membership. This often requires modifications to the algorithm to correct imbalances attributed to differences in group attributes.
- **Disparate Impact**: This measures the proportionate outcomes for different groups, ensuring that no group is disproportionately impacted by AI-driven decisions. A lower score in disparate impact indicates that the AI system is fairer.
- **Calibration**: Calibration relates to the accuracy of predictions among different groups. An AI system is well-calibrated if it predicts outcomes

equally well (or poorly) for all groups, maintaining consistency whether the predictions are correct or incorrect.

- **Ethical Considerations and Frameworks**: The development of fair AI systems is underpinned by ethical considerations that guide the design, implementation, and deployment of these technologies. Several frameworks have been proposed to embed ethics in AI:
  - **The Asilomar AI Principles**: These principles include guidelines for research priorities, ethics and values, and longer-term issues, focusing on developing AI in ways that benefit society and minimize risks.
  - **IEEE Ethically Aligned Design**: This framework provides a comprehensive set of recommendations to ensure that AI systems align with ethical principles and human values, promoting human well-being, data agency, and algorithmic transparency.
  - **EU Guidelines on AI Ethics**: The European Union's guidelines emphasize trustworthy AI, defined by its lawfulness, ethicality, and robustness. These guidelines focus on ensuring respect for human autonomy, prevention of harm, fairness, and explicability.

These measures and frameworks are essential for ensuring that AI systems operate fairly and ethically, providing equal treatment and opportunities across all demographic groups.

Implementing fairness in AI systems

With the rise of AI, healthcare is on the brink of a revolution that promises sweeping changes in the care, diagnosis, and management of patients, potentially redefining medical practice. The integration of AI technologies enhances diagnostic accuracy, allows for highly personalized treatment plans, improves patient monitoring, and boosts operational efficiencies. However, incorporating AI into healthcare comes with significant challenges, including fairness and bias, which pose both technical and ethical dilemmas.

Bias in AI can lead to inconsistencies in the quality of care delivered across different demographic groups, exacerbating existing healthcare disparities. Unchecked biases can diminish trust in healthcare systems, reduce patient satisfaction, and compromise the effectiveness of AI healthcare solutions. Addressing these issues is not just a technical necessity but also a moral imperative to uphold justice and equity in medical services.

This chapter delves into the multifaceted issue of bias in healthcare AI, exploring its origins, manifestations, impacts on patient outcomes, and public trust. It also examines various methodologies aimed at mitigating these biases and discusses proactive strategies to integrate fairness throughout the lifecycle of AI healthcare systems, from design to deployment and monitoring.

The cyclical, iterative, and continuous process of maintaining AI integrity includes.

1. **Data Collection**: Gathering a diverse range of data to ensure the AI system's training data is representative of the entire population it serves.
2. **Data Processing**: Addressing issues such as missing data or samples that aren't proportionally representative of the population to identify and correct biases that could affect the AI system's fairness.
3. **Algorithm Design**: Developing AI algorithms that are aware of fairness concerns and include mechanisms to detect and mitigate bias.
4. **Validation and Testing**: Rigorously testing the AI system across various demographic groups to ensure it does not favor one group over another.
5. **Deployment Monitoring**: Continuously monitoring the AI system postdeployment to ensure it operates fairly in real-world applications and to identify any new biases that may emerge.
6. **Feedback Loop**: Using feedback from the monitoring stage to refine and improve the AI system, ensuring it evolves and adapts to maintain fairness over time.

This comprehensive approach aims to ensure the principles of equity and justice are embedded in the rapidly advancing field of healthcare technology.

Conclusion

In short, AI use in health care has huge potential to improve the quality of care for patients and appropriateness of treatment, with advantages for operational efficiency. Realizing this promise will require overcoming significant challenges that—particularly in the areas of fairness and bias—can have outsized consequences for the quality of care across demographic groups. These issues shall be important not only for the purposes of harnessing huge potentials that AI has in healthcare but also fundamentally for ensuring equity and justice in medical practice.

Although there are many problems, remedies for biased AI demand multidisciplinary efforts that cut across providers, developers, and policymakers in the health field. In addition, this requires strong measures to ensure diversified and representative data collection, rigorous testing of the algorithms, and continuous monitoring and adaptation of the AI systems in accordance with the feedback and the changing standards of ethics. This multidimensional approach to fairness in the healthcare sector will ensure that AI technologies deliver equitable health outcomes and retain public trust.

In addition, the fast development of AI requires continuous updates of protocols for fairness, reflecting the newest changes and findings. Just as AI technologies advance, so must our approaches in assuring that such tools advance equal health for all patients and do so without bias and discrimination. And leading with fairness in AI, the health industry might present a leading example in using these very advanced technologies to ensure not only improved health outcomes but also increased justice and equity.

References

Char, D.S., Shah, N.H., Magnus, D., 2020. Implementing machine learning in health care — addressing ethical challenges. N. Engl. J. Med. 378 (11), 981–983.

Chen, I.Y., Szolovits, P., Ghassemi, M., 2018. Can AI help reduce disparities in general medical and mental health care? AMA J. Ethics 20 (2), 167–179.

Esteva, A., Robicquet, A., Ramsundar, B., Kuleshov, V., DePristo, M., Chou, K., Cui, C., Corrado, G., Thrun, S., Dean, J., 2019. A guide to deep learning in healthcare. Nat. Med. 25 (1), 24–29. https://doi.org/10.1038/s41591-018-0316-z.

Ghassemi, M., Naumann, T., Schulam, P., Beam, A.L., Ranganath, R., 2018. Opportunities and challenges in developing machine learning models using electronic health records data: a systematic review. J. Am. Med. Inf. Assoc. 25 (12), 1419–1428.

Gianfrancesco, M.A., Tamang, S., Yazdany, J., Schmajuk, G., 2018. Potential biases in machine learning algorithms using electronic health record data. JAMA Intern. Med. 178 (11), 1544–1547. https://doi.org/10.1001/jamainternmed.2018.3763.

Joyce, N.C., Topol, E.J., 2019. A new era for the use of artificial intelligence in personalized medicine. JAMA Netw. Open 2 (7).

Obermeyer, Z., Powers, B., Vogeli, C., Mullainathan, S., 2019. Dissecting racial bias in an algorithm used to manage the health of populations. Science 366 (6464), 447–453. https://doi.org/10.1126/science.aax2342.

Rajkomar, A., Hardt, M., Howell, M.D., Corrado, G., Chin, M.H., 2018. Ensuring fairness in machine learning to advance health equity. Ann. Intern. Med. 169 (12), 866–872. https://doi.org/10.7326/M18-1990.

Vyas, D.A., Eisenstein, L.G., Jones, D.S., 2020. Hidden in plain sight — reconsidering the use of race correction in clinical algorithms. N. Engl. J. Med. 383, 874–882.

Further reading

Angwin, J., Larson, J., Mattu, S., Kirchner, L., 2016. Machine bias. ProPublica.

Corbett-Davies, S., Goel, S., 2018. The Measure and Mismeasure of Fairness: A Critical Review of Fair Machine Learning arXiv preprint arXiv:1808.00023.

Crawford, K., Calo, R., 2016. There is a blind spot in AI research. Nature 538 (7625), 311–313.

Dwork, C., Hardt, M., Pitassi, T., Reingold, O., Zemel, R., 2012. Fairness through awareness. Proceedings of the 3rd Innovations in Theoretical Computer Science Conference.

Eubanks, V., 2018. Automating Inequality: How High-Tech Tools Profile, Police, and Punish the Poor. St. Martin's Press.

Goodman, B., Flaxman, S., 2017. European Union regulations on algorithmic decision-making and a "right to explanation". AI Mag. 38 (3), 50–57.

Kallus, N., Zhou, A., 2018. Residual unfairness in fair machine learning from prejudiced data. In: Proceedings of the 35th International Conference on Machine Learning. ICML 2018.

Raji, I.D., Buolamwini, J., 2019. Actionable auditing: investigating the impact of publicly naming biased performance results of commercial AI products. In: AIES '19: Proceedings of the 2019 AAAI/ACM Conference on AI, Ethics, and Society.

Veinot, T.C., Mitchell, H., Ancker, J.S., 2018. Good intentions are not enough: how informatics interventions can worsen inequality. J. Am. Med. Inf. Assoc. 25 (8), 1080–1088.

Wiens, J., Saria, S., Sendak, M., Ghassemi, M., Liu, V.X., Doshi-Velez, F., Jung, K., 2019. Do no harm: a roadmap for responsible machine learning for health care. Nat. Med. 25 (9), 1337–1340.

Zou, J., Schiebinger, L., 2018. AI can be sexist and racist — it's time to make it fair. Nature 559, 324–326.

CHAPTER 6

AI enhanced healthcare: Opportunities, challenges, ethical considerations, and future risk

Soma Gupta, Satarupa Mohanty, Dayal Kumar Behera and Soumya Ranjan Nayak
School of Computer Engineering, KIIT Deemed to Be University, Bhubaneswar, Odisha, India

Introduction

Worldwide healthcare organizations have significant challenges in achieving objectives such as improving the population's overall health, enhancing the patient and caregiver experience, and lowering the steadily rising cost of care. Healthcare is at present utilizing AI and related technologies, which are becoming more and more prevalent in industry and society. These advancements have the potential to change a variety of patient care features as well as the pharmaceutical organization and administrative procedures. AIaugmented healthcare systems have the potential to completely change healthcare delivery paradigms due to the growing availability of numerous kinds of data and the developments in mobile, Internet of Things, computational power, and data security. Finally, healthcare is seeing a growth in AI and Machine Learning investments.

Several studies have been carried out that indicate AI has the potential to perform crucial medical tasks such as identifying diseases equally well as well as possibly better than people. At the moment, algorithms are surpassing radiologists in the detection of hazardous malignancies and in guiding the formation of groups for costly clinical investigations. However, we anticipate that it will be some time before AI completely replaces human beings in medical procedures or early identification and diagnostic purposes. It was developed in several methods to better precisely, consistently, and promptly detect disorders such as cancer.

Responsible and Explainable Artificial Intelligence in Healthcare
ISBN 978-0-443-24788-0
https://doi.org/10.1016/B978-0-443-24788-0.00006-6

It compares information from certain patients, including images of a sizable amount of information from other patients. A relationship is found and the independent learning system suggests a diagnosis. Google's Deep Mind Health Technology is an illustration of this since it combines the use of machine learning with a neuroscience system to create an AI model that mimics the human brain and assists physicians in diagnosis and decision-making.

Numerous risks and difficulties exist, including the potential for patient harm from system flaws, the potential for patient privacy breaches when using data, and the use of artificial intelligence to conclude. AI must be capable of assisting individuals in preserving their health for public health, and attempt to give more control over their well-being and health.

The remaining sections of the paper are organized as follows. Section 2 describes Related work. Section 3 covers AI in healthcare. Section 4 discusses Applying AI in healthcare. Section 5 discusses Ethical principles in AI-powered diagnosis and treatment. Sections 6 and 7 describe the Challenges of AI in healthcare and Future of AI in health care. The conclusion is provided in section Section 8 Conclusion.

Related work

The privacy of patients is an important concern when using AI in the healthcare sector. The very sensitive Personally Identifiable Information (PII) that makes up patient data, including medical histories, identifying details, and payment information is protected by GDPR and HIPAA regulations.

To protect healthcare organizations, providers and patient's privacy and security manufacturers must now more than ever integrate comprehensive security measures during the design process of medical devices built on AI.

This systematic literature review follows three main approaches to healthcare systems that rely on natural language processing, computer vision, and acoustic AI. They discovered that (1) There are certain types of defense mechanisms for adversarial attacks on AI and AI-based adversarial techniques are vulnerable to numerous attacks. (2) Privacy and security concerns in the healthcare industry can be addressed by Blockchain (3) Blockchain can enable user provenance and medical data verification. (4) Distributed learning on heterogeneous medical data can be protected by blockchain. (5) Blockchain technology can be used to address problems with healthcare systems such as single points of failure and lack of transparency. However, it has been noted that the study is still in its early stages (Shinde et al., 2022).

The emergence of artificial intelligence and IoT has recently opened up the potential for expanding medical and healthcare services of patients, cutting costs, and enhancing the wellness of communities, which were previously unimagined opportunities. The storage and secure flow of data, however, remains a critical problem for the current healthcare management system. This research presents a revolutionary Lionized Remora optimization serpent (LRO-S) encryption method that secures critical data while minimizing violations of privacy and cyber-attacks by unauthorized users and hackers (Almalawi et al., 2023).

The main goal is to maximize the use of available resources and technologies to improve the quality of life. These resources and innovations include support and efficiency in various domains including healthcare, agriculture, transportation, agriculture, and smart grid. To tackle such challenges, different cutting-edge technologies blockchain combined with AI and ML can be utilized to effectively handle the previously addressed data security and protection problems (Khurana, 2023a).

Smart healthcare systems are vulnerable to a variety of security risks and threats, including network assaults, system-level attacks, hardware and software-based attacks, and attacks that could endanger the lives of patients. The majority of healthcare-based surveys on security focused largely on security issues and did not incorporate the integration of cutting-edge technology to address security concerns, exposing a research gap, according to a review of the available research. In this study, a thorough analysis of many most recent security concerns and their mitigation strategies in the context of smart healthcare is performed. Furthermore, as a case study, a secure Blockchain and Artificial Intelligence-based architecture is offered to investigate malware and network threats on wearables. A range of performance metrics, including dynamic malware analysis, blockchain scalability, and accuracy, are used to evaluate the proposed architecture (Alabdulatif et al., 2022).

The purpose of the research was to ascertain the prospective benefits and hazards of AI in medicine. Google Scholar, Web of Science, and EBSCO host-three databases were reviewed for works that dealt with using AI to enhance health services. Despite numerous studies conducted across the globe, only a small number of AI-based solutions have actually reached clinics. Nonstandardized medical records, an abundance of maintained records, and rigorous ethical and legal rules protecting patient privacy are major obstacles to the wide acceptance of medically verified AI applications (Sunarti et al., 2021).

The creation of privacy-preserving methods and resolving the problems preventing AI use in a healthcare setting have received a lot of attention in the literature. This study complies with the most recent methods for protecting privacy in AI-based healthcare applications to achieve this (Khalid et al., 2023).

About TREs (Trusted Research Environments) in the UK, this study addressed advancements in medical data and machine learning research as well as policies in this article. It is discovered how TREs are positioned at the crossroads of growing demand for information-driven and machine-learning solutions for healthcare. In conclusion, it emphasize three significant legal and ethical challenges to maintain a safe environment for medical and healthcare data that serve the people (1) the possibility of ML models releasing sensitive data (2) the evolving nature of machine learning models, and (3) how societal benefits may be (re)conceived in this context (Kerasidou et al., 2023).

AI in healthcare

AI is the engineering and scientific method of creating intelligent computers using algorithms or a collection of rules that the computer to replicate cognitive abilities such as acquiring knowledge and solving problems. The term "AI" refers to sever branches such as machine and deep learning that when applied independently or collectively, bring intelligence to applications rather than being one complete, ubiquitous technology, which is shown in Fig. 6.1.

In healthcare ML, NLP, Deep Learning, Expert System are widely used AI tools.

Although most of these technologies are instantly useful in the healthcare sector, the specific tasks and processes they support vary substantially. Here are some AI technologies that are crucial to the healthcare industry (Rahman et al., 2021).

Machine learning (ML)

ML is a branch of AI that enables machines to learn from experience and progress without being programmed explicitly. It is a subset of Artificial Intelligence. It is the study of making machines to learn and develop their programs, which will enable them to behave and make decisions more as

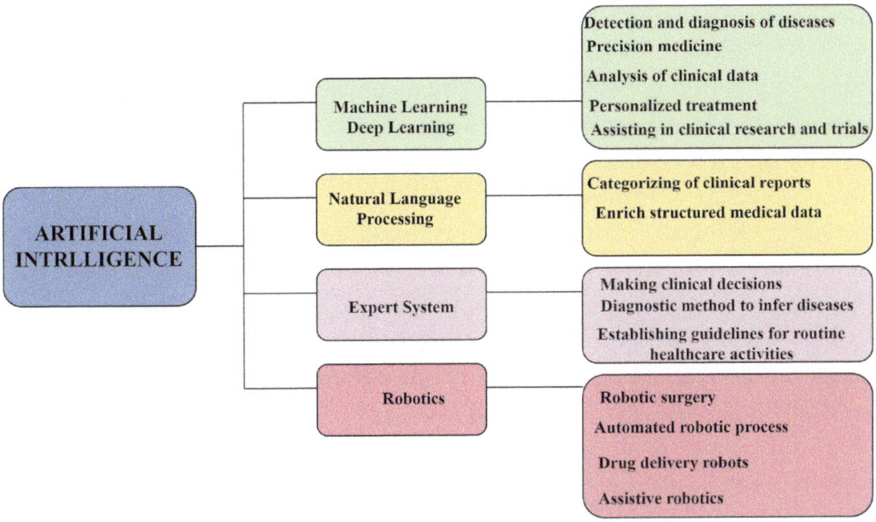

Figure 6.1 AI in healthcare few artificial intelligence (AI) tools in healthcare.

humans would. With little to no explicit programming and minimum human intervention this is accomplished. Based on what the machines learn during the process the learning process is automated and improved.

Machine learning algorithms could considerably aid the healthcare business by assisting in the interpretation of the massive volumes of data related to healthcare generated every day. Using ML techniques, we can detect trends and insights in medical data that would be hard to detect manually.

Medical professionals can utilize machine learning for enhanced diagnostic tools for analyzing images from medical procedures. For instance, machine learning can be used to search for patterns that signal a specific condition in medical imaging (such as X-rays or MRI scans). This type of Machine learning algorithm could potentially help doctors make quicker, more accurate diagnoses leading to improved patient outcomes.

A subset of machine learning called "Deep Learning" uses artificial neural networks as its foundation. It can discover links and patterns in data. Every component of deep learning does not need to be explicitly programmed. It is becoming increasingly popular because of advancements in processing power and the accessibility of enormous datasets. Deep Learning is a type of AI and machine learning that mimics how people acquire specific types of knowledge. Similar to how millions of interconnected neurons in human brains collaborate to learn information, deep learning uses neural networks built from many layers of working software nodes. A substantial amount of labeled data and neural network designs are used to train deep learning models. A machine can learn from examples in deep learning. In Deep Learning models, data is filtered via a cascade of several layers with each layer utilizing the output from the layer before it to decide upon the result. Deep learning models improve in accuracy as more data is analyzed, essentially learning from earlier outcomes to increase their ability to identify correlations and connections. The study of brain Magnetic Resonance Imaging (MRI) scans to predict Alzheimer's disease and its variant was one of the first applications of deep learning to clinical data in image processing (Liu et al., 2014). DL has emerged as the predominant method in AI for advancing fields such as image and speech recognition. The study discusses the potential benefits and constraints of deep learning and IoT for healthcare and explains the various deep learning methods employed by researchers and practitioners globally in connection with IoT (Singh et al., 2021).

Depending on the specific application, machine learning models in healthcare can be quite complex. Fig. 6.2 shows the various steps involved in developing machine learning models for healthcare.

ML framework is used for faster and easier building and deployment of machine learning models.

Natural language processing (NLP)

NLP is an area of study within AI. This technology analyzes the speech that the user provides, breaks down it for accurate understanding, and processes it as necessary. This is a relatively new and successful strategy, and as a result, there is a huge demand for it in the market at present. In the rapidly evolving field of NLP, various innovations have already been made feasible, such as interaction with smart devices and interactive human discussions.

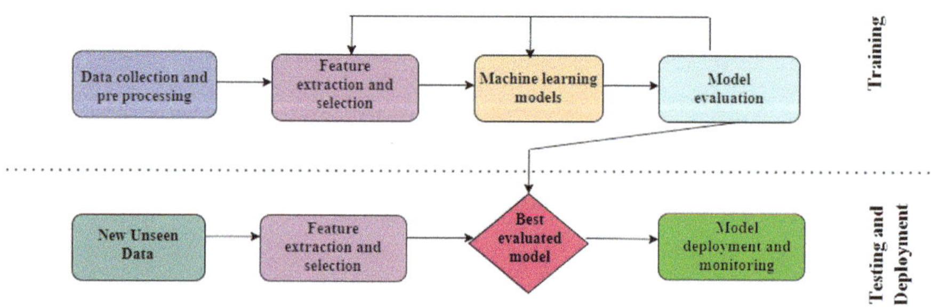

Figure 6.2 ML based system framework conceptual ML framework.

Semantic NLP and Statistical NLP are the two primary approaches to it. NLP transforms language that humans use into an organized and understandable format so that computers can use it for a variety of computational analyses. The ability to understand human language is not inherent to computers. Without additional instructions, they are unable to analyze human sentences to recognize sentence components such as nouns, verbs, and adjectives, much less identify the noun that an adjective refers to or understands the idea of negation. As a result, the data appears to be in an unstructured format. NLP makes an effort to compensate up for these shortcomings by processing text through an NLP engine, a program intended to codify human language. The output of the NLP engine is a collection of structured data that accurately reflects the contents of the provided text and is understandable by and usable by the computers.

The use of NLP will significantly speed up the procedures for making decisions in the field of healthcare. However, the true benefits of creating effective algorithms will be greatly influenced by the quality of the data they collect and maintain. The quicker decision-making process will allow specialists to concentrate on providing patients with value-added care (Jain and Prajapati, 2021). The enormous amount of data generated every day and the finite cognitive capacity of the human mind may be bridged through the use of natural language processing methods (Nair et al., 2019). NLP has practically endless potential to transform electronic health records from the most advanced application in precision medicine to the basic process of coding a claim for billing and reimbursement. The conversion of data representing the patient's problems, medications, allergies, and previous medical and surgical history into data that can be codified and stored in an electronic medical record is a perfect instance of NLP being used in the clinical setting. Other applications of NLP in health care include codifying the information from a radiologist's transcription of a mammogram interpretation into codified registry data and adding noncodified data from other hospital's electronic documents to patient's current electronic medical records. NLP can be used to transform delivered speech into organized and codified information that is useful by a computer application when combined with the usage of speech-to-text programs that converts spoken words into text.

Expert system

An expert system is a trustworthy and interactive computer-based decision-making system. To address complex decision-making challenges, it employs heuristics and facts. It is regarded as the pinnacle of human intelligence and expertise. The AI Expert System is capable of solving a wide range of issues that would ordinarily require human experts. It is based on data that was obtained from a reliable source.

An area of the healthcare sector where knowledge-based technologies and Expert Systems have become prevalent is managed/care utilization. This unique ES technology application has the potential to change the norms of medical practice by incorporating "medical necessity criteria" into the knowledge base and sharing significant results from studies about treatment outcomes (McCauley and Ala, 1992).

Computer systems that mimic the diagnostic decision-making abilities of human experts are known as diagnostic expert base systems, for infectious disease Mycin is a notable system, and for general internal medicine, Internist-1, QMP, and Dxplain are noteworthy systems. To construct a set of principles in a particular knowledge domain, expert systems require the assistance of human experts and knowledge engineers. Up to a point, they are easy to use and understand. The situation begins to unravel, though, when there are too many regulations and they begin to conflict. In addition, altering the rules may be complex and time-consuming if the knowledge domain changes. In the healthcare industry, more data-driven and machine-learning-based approaches are gradually replacing them.

Robots and robotic process

Physical robots are currently well known since over 200,000 of them are installed annually worldwide. In places such as factories and warehouses, they do preset tasks such as lifting, moving, welding, assembling items, and delivering materials in healthcare facilities.

The science of artificial intelligence and machine learning is as prevalent as electricity in the modern world, therefore its use in robotics is also growing more important. Precision machine learning techniques are being used to train robots and boost their accuracy. Through the use of artificial intelligence, robots may learn how to carry out a variety of tasks, including grasping objects and comprehending spatial relationships and other tasks requiring computer vision and motion control.

As technology develops, robots are used to assist surgeons during complex and delicate surgeries as well as to care for patients in hospitals and rehabilitation centers. Robots improve ergonomics and convenience of surgeons in operating rooms by enabling more procedures to be performed at once, reducing surgeon fatigue. Additionally, robots assist in removing human mistakes from high-risk procedures by analyzing vast volumes of data and finding patterns that may assist researchers and medical professionals in better-understanding illnesses such as cancer, liver, kidney, and heart problems.

According to this review, robots have carried out 10 major tasks in a variety of medical environments. The most common duties were surgical and mobility and rehabilitation, Although the surgery theatre and the rehabilitation unit were the primary study sites for robots, other places of study included the hospital ward and the inpatient pharmacy. COVID-19 shows how healthcare demands are always changing and how robots could help with adaptation. Infrastructure systems will need to advance to support more telepresence in the future (Morgan et al., 2022).

Applying AI in healthcare

Future healthcare technologies will be greatly impacted by AI. Nevertheless, the early attempts at offering advice on a proper diagnosis and course of treatment have been challenging. Eventually, it's predicted that AI will excel in that field as well. Considering the fast development of AI in imaging, it appears plausible that machines will eventually evaluate the majority of pathology and radiology images. Actions such as patient interaction and taking clinical notes already involve recognition of text and speech, and their use will increase. Fig. 6.3 describes applications of AI in various aspects of healthcare.

AI plays an important role in diagnosis, robot assisted surgery, administrative applications and personalized healthcare plans.

This study examines several techniques and algorithms for healthcare diagnosis and Imaging using cutting edge tools such as deep learning and machine learning (Rakhra et al., 2021).

Due to the digitization revolution, the healthcare sector is currently overwhelmed with data. Electronic health records are the first of several data streams that are included in the healthcare system (EHR). The study objective is to set the groundwork for healthcare system adoption of the elements necessary to do comprehensive analytics to expand and create

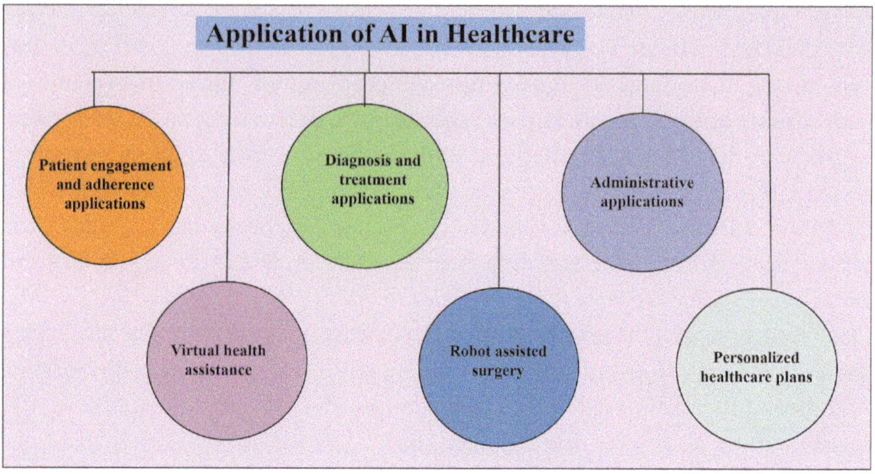

Figure 6.3 Role of AI in healthcare application of AI in various aspects of health care.

algorithms to streamline daily operations and to improve and expand human life (Saravanan et al., 2021). In the field of medical car, fog computing offers a practical and efficient means of improving the delivery of outstanding services. This research offers a novel fog computing-based platform for remote health monitoring (Khurana, 2023b).

Patient engagement and adherence applications

Engagement of patients and adherence, generally characterized as the "final mile" problem in healthcare, are one of the biggest barriers between poor and beneficial healthcare results (Kuwaiti, 2023). Patients who are more proactive about their health and care have better application, financial outcomes, and member satisfaction. Big Data and AI are increasingly being used to address these challenges. Healthcare organizations and clinicals frequently rely on their clinical knowledge to establish a treatment plan that will enhance the well-being of a chronic or acute patient. Nonetheless, it usually makes little difference if someone doesn't follow prescribed medication, lose weight, schedule a follow-up appointment, or follow a treatment plan.

Noncompliance is a major issue that occurs when someone fails to adhere to a certain course of treatment or utilize the prescribed medications. Another area of expanding interest in health care is effectively developing the "choice architecture" to influence patient behavior in a more predictive manner using empirical evidence. Information from ERH systems, biosensors, watches, smartphones, conversational interfaces, and other devices may be used by software to customize suggestions by comparing patient data with other effective treatment methods. This recommendation may be given to doctors, patients, nurses, call center representatives, or care delivery coordinators.

In terms of patient engagement and adherence, AI is becoming more and more important. It's generally accepted that patients' health improves as they become more actively involved in their care. AI-enabled applications can help patients stay on track with their treatment regimens and improve health outcomes by offering personalized recommendations and reminders.

AI can aid in the early identification of possible nonadherence concerns in addition to boosting patient involvement. AI-powered applications can provide insights that enable care teams to intervene before nonadherence becomes a major issue by evaluating a patient's behaviour. AI can be used by healthcare practitioners to improve patient engagement and adherence, improve health outcomes, and boost the efficiency of the healthcare system.

Patients who are actively involved in their healthcare will likely have better usage, cost, and patient experience outcomes. When a patient neglects to make the necessary behavioural changes, such as managing a healthy weight, scheduling appointments, and adhering to the prescribed therapy, it generally makes no difference.

Diagnosis and treatment applications

AI may also offer suggestions and support for the opinions of medical professionals based on assessments of vast volumes of healthcare data, such as electronic health records (EHRs), illness data, diagnostic imaging reports from healthcare providers, and more. This does not imply that AI can take the role of doctors in the detection and treatment of diseases and other ailments. This could facilitate and expedite decision-making during diagnosis and treatment.

There are various ways AI can assist doctors in providing better care. Following the evaluation of enormous volumes of patient and treatment data, to discover what has previously worked for similar patient cohorts, treatment plans, and other follow-up measures can be personalized more successfully. This can aid in addressing the persistent issue of patient noncompliance together with better messaging. The development of medications and better drug targeting to increase efficacy and decrease adverse drug reactions (ADRs) are further treatment uses for AI in healthcare.

We found a significant number of systematic reviews providing research in which AI was one of the diagnostic or therapeutic tools for a variety of disorders. As a test case, it can be demonstrated that AI seems to have high accuracy for the diagnosis of particular medical states in the context of digestive system disease. The methodical efficacy of SRS of AI in the identification and management of diseases, on the other hand, needs to be enhanced, particularly in the areas of publication bias and scientific assessment of the quality of the associated research (Wu et al., 2022).

Machine learning and deep learning techniques in artificial intelligence are frequently employed for disease detection. This article provides a thorough investigation of methods using artificial intelligence for identifying a variety of diseases such as Alzheimer's, cancer, diabetes, cardiovascular disease, TB, stroke, high blood pressure, skin, liver, and other conditions. They conducted an in-depth study, utilizing feature extraction and classification from medical imaging datasets to provide predictions (Kumar et al., 2023).

This work proposes an automated ensemble deep-learning model for Diabetic Retinopathy detection and categorization. Better forecasts and performance are possible with the amalgamation of deep learning models than with the contributions of individual models (Mondal et al., 2023).

Administrative applications

Administrative costs are estimated to make up about 15%–25% of total healthcare expenses. Tools that improve and streamline administration can be beneficial for insurance, businesses, payers, and providers. The administrative applications in healthcare are countless.

Although the implementation of AI in this field is less likely to revolutionize than in healthcare, it can nonetheless achieve tremendous efficiency. In the healthcare sector, for instance, a typical US nurse devotes 25% of their time at work to administrative and regulatory duties. The most likely technology to be useful for this goal is RPA. Among the healthcare applications it can be used for are claims processing, clinical recording, managing the revenue cycle, and healthcare records administration.

However, as healthcare fraud can take place on many different levels and be perpetrated by several parties, discovering and minimizing fraud might provide the most beneficial outcome. In rare cases, fraud may result in surgeons undertaking unnecessary surgeries to enhance insurance reimbursements or insurers being billed for never-off treatments. Additionally insures might be charged for test kits or defective devices.

AI has the potential to prevent fraud before it occurs. Health insurers can utilize algorithms to identify anomalous transactions in the same way that banks do frequently. The management of claims and payments may take advantage of another AI approach called machine learning, which can be used to probabilistically match data around several database systems.

Insurance firms must ensure that millions of claims are accurate. All parties benefit from correctly recognizing, analyzing, and resolving coding errors and false claims, which saves money, time, and effort.

Through data matching and claim audits, inaccuracies in claims that get missed can represent enormous profit potential. Healthcare workers and staff devote more time to managerial duties and there is insufficient information on work patterns to help with resource allocation. Historically, it has been difficult to compile accurate data on work done in medical environments.

This article presents an approach for enhancing knowledge of managerial tasks and completion of task networks, thereby lessening the burden of administrative chores by using AI to gain knowledge and improve system behaviour (Glover et al., 2022).

Much focus has been placed on the risk that AI would result in job automation and significant labor displacement. According to investigation, the expense of automation technologies, the size, and cost of the labor market, the benefits of automation other than simple labor substitution, legislative, social acceptance, and the viability of some job automation could all minimize the loss of employment. These factors might prevent employment losses from exceeding 5%. As of yet, AI has not resulted in the loss of any jobs of healthcare providers. The lack of effect on employment has primarily been attributed to the industry's slow adoption of AI and the difficulty in integrating it into clinical procedures and EHR systems. Jobs that deal with digital change rather than those that involve direct patient contact, such as radiology and pathology, are the ones in healthcare that are most likely to be automated. AI adoption in fields such as radiology and pathology will likely be gradual. Technologies such as deep learning are enhancing our capacity to analyze and categorize photos.

Cyclic data processing is required to produce meaningful, actionable judgments in the healthcare industry. Healthcare workers' occupational stress has increased due to the rapid growth of clinical data, which affects their ability to provide high-quality services. Health institutions must radically rethink their strategies if they want to make sure that personnel are happy and actively encouraged in their work. The performance of providers may be improved using artificial intelligence (AI). This article performs a review of the literature to identify areas where AI has the potential to alter the job of healthcare professionals. To fully utilize AI, policymakers, corporate leaders, medical experts, and patients must address a new set of concerns (Hazarika, 2020).

Ethical principles in AI-powered diagnosis and treatment

The ethical standards and values that all significant stakeholders uphold and adhere to must guide the development and application of AI technology in healthcare. Various datasets and approaches such as supervised, semi-supervised, and unsupervised learning, are used in AI technology. Health professionals and researchers have been advised to proceed cautiously despite the promising nature of these "machine-driven" analytical tools due to their intricacy. In contrast to other AI sectors, AI for health directly affects human existence and can potentially harm every facet of patient's lives. As a result, before these algorithms can become routine in the health industry, a cautious but unobtrusive and ethical approach is required.

Concerns about AI in healthcare include discrimination and bias, privacy and monitoring and, the utilization of human judgement. Healthcare errors can have severe consequences for patients since there is always the potential for inaccuracies and data breaches wherever technology exists. It's an important topic that must be researched because there are no defined limitations on the ethical and legal challenges associated with AI and its role in healthcare. The potential for accuracy issues and data breaches is raised by the use of AI in healthcare. Digital health records can be used to optimize clinical care, improve healthcare quality, and conduct scientific research, yet there is a risk that the data will be corrupted and exploited. Additional ethical considerations include who owns a person's medical records and patient history, who will have access to them, and when and if consent is required.

Numerous ethical considerations involved with the use of AI in healthcare is described in Fig. 6.4. Transparency, accountability, consent, and privacy are issues when employing intelligent computers to assist in healthcare decision-making since almost all healthcare choices have

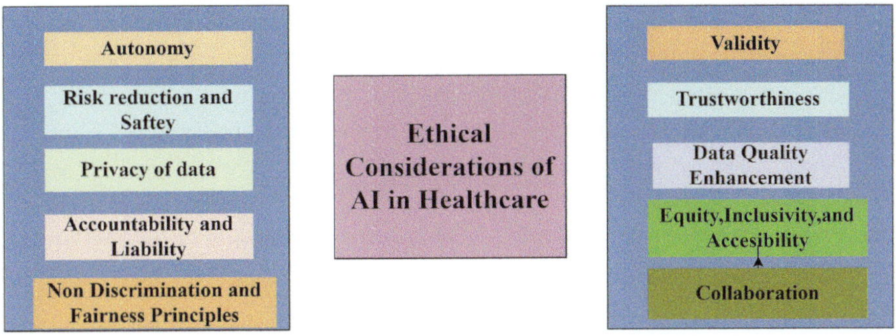

Figure 6.4 Ethics of AI in healthcare various ethical considerations of AI in healthcare.

historically been made by people. In the context of modern technology, transparency might be the hardest problem to solve. Many AI systems, particularly deep learning algorithms used for analysis of image are hard for humans to understand or interpret.

Ethical Considerations in AI includes concerns with patients rights, privacy, bias and discrimination, accountability.

Even though there are no particular law addressing them, the application of AI in healthcare may raise ethical and legal concerns. The necessity of algorithmic transparency, privacy, and protection for all associated beneficiaries, as well as cyber security of related vulnerabilities, is emphasized in this review (Naik et al., 2022).

While treating and diagnosing patients, artificially intelligent machines will undoubtedly make mistakes, and it might be challenging to hold them accountable. Additionally, patients may occasionally obtain medical data from Artificial Intelligence system that they would prefer to receive from a physician. Healthcare based machine learning algorithms may also be susceptible to bias in algorithms, which predicts an increased risk of illness based on gender or ethnicity even when those are not the true causes.

The use of AI for medical purposes is expected to result in several technological, ethical, medicinal, and professional advancements. It is critical for healthcare organizations, and also for governmental and regulatory organizations, to establish frameworks to track significant issues, respond responsibly, and develop governance mechanisms. Any negative impacts will be lessened as a result. Because it has one of the most significant effects on humans, this technology will necessitate ongoing oversight and cautious policymaking for many years.

Concerning AI for healthcare, the ethical principles address particular challenges. It is anticipated that these patient-centered principles will direct all parties involved in the development and application of trustworthy and responsible AI for health care. These principles are as follows.

Autonomy

The potential for the system to operate independently and undermine individual autonomy exists when AI is used in the healthcare industry. The deployment of AI technology in healthcare may place decision-making in the hands of machines. Under no situation should AI based technology interfere with the autonomy of the patient.

Clinicians and AI technology may make different medical conclusions, making it difficult for the individual using it to decide whom to believe the AI system or the clinician. Both alternatives ought to be given to patients in such cases. Before implementing any AI technology in healthcare, individuals need to be fully aware of the advantages as well as any associated social, psychological, and medical risks. The choice of whether or not to adopt AI technologies must be entirely up to the patients. Human values and morals should be effectively and transparently monitored at every stage of the development and application of AI.

Consent may be refused by the patient or participant. The use of such AI technology should not be subject to compulsion from government agencies, sponsors, medical professionals and, other stakeholders. A patient's autonomy and the relationship between them and their physician may suffer if AI technologies are used excessively for diagnosis and treatment. Therefore, to promote participant autonomy, rules and regulations should be adopted by all stakeholders.

Risk reduction and safety

Before an AI-based system is widely used, it must be proven that it will run securely and reliably. All stakeholders participating in the development and implementation of AI technology bear responsibility for assuring participants' safety. Prioritizing the safety, well-being, and dignity of patients and participants is vital. The kind of application and implementation method selected will determine the risk involved with AI technology and approaches in medical research and patient care.

Some risk reduction and safety considerations are listed below.

1. To prevent unintentional or deliberate misuse, a robust set of control mechanisms is required.
2. Every AI based technology and algorithms needs to be thoroughly evaluated in the environments in which they are intended to be applied. It is necessary to evaluate the algorithm's effectiveness across racial and ethnic categories, age ranges, socioeconomic strata, and other relevant human characteristics.
3. The Ethical Committee (EC) and other relevant parties must guarantee that the benefit-risk balance is beneficial. The advantage must outweigh the risk.
4. Technologies based on AI should be created strictly adhering to the fundamental legal, ethical and, data protection needs of the nation.
5. A reliable, well-defined procedure should be in place to routinely assess the capabilities, vulnerabilities, and security needs of AI technology.

Privacy of data

At every stage of development and deployment, AI-based technology should guarantee privacy and safeguard personal data. For AI to be used effectively and widely, it is imperative that all stakeholders those receiving healthcare continue to have faith in the safe and secure handling of their personal data. Preventing unauthorized access, modification, and loss of personal data should be the aim of data privacy. These procedures are essential to the healthcare industry because medical records are sensitive even inadvertently, and could cause people harm or discrimination. Unless it's necessary for medical or research purposes, specific patient information should be anonymized. Before any sort of data sharing, all algorithms that handle patient data must provide sufficient anonymization.

Concerns about patient privacy are also raised by AI-powered diagnostics. Patient data is frequently needed to train AI systems, and there is a chance that this data will be stolen or used without the consent of the patient. To preserve patient privacy, it is critical to develop strong data protection rules that prioritize patient consent and confidentiality.

Some key points about data privacy are highlighted below.

1. Users should have access to the data that has been acquired about them in to build and construct AI systems for healthcare. Users should have the ability to access, alter, and delete such data from AI technology at any time.

2. Individuals using it must be properly informed about the privacy precautions in place. The kind of data that is collected and its intended use be it storing, analyzing or creating AI algorithms must be thoroughly explained to them.
3. Before executing a predictive algorithm on participants/patients, consent must be obtained.
4. Before adopting AI for general usage, relevant authorities must conduct an impact assessment. It should concentrate on critical issues such as human rights, privacy, and ethical ideals.
5. Repurposing excess data without the patient's or participant's consent is unethical. If not done earlier, storing excess data for future usage may necessitate extra consent.

Accountability and liability

Patients and professionals may not fully comprehend how AI algorithms make diagnosis, leading to suspicion and skepticism. It is critical to involve patients and physicians in the development and implementation of AI-powered diagnostic tools to promote transparency and accountability. This includes explaining how the AI algorithms function in detail and ensuring that procedures are in place to monitor and assess their performance.

A person or an organization is said to be accountable if they are willing to take ownership of their actions, account for their activities, and transparently report the outcomes. AI technologies designed for use in the health sector must be prepared to be scrutinized by relevant authorities at any moment. To achieve optimal performance, AI technology must be subjected to regular internal and external assessments. The public must have access to these audit reports. The creators of AI must permit independent review and analysis.

Clinical ethics, research guidelines, and legal constraints may be unfamiliar to AI innovators. Therefore, it is essential to involve members of the healthcare sector at every stage of the creation and application of AI-based tools and technologies.

Automation is by far the most evident benefit of AI-based systems. Algorithms for machine-assisted decision making are employed in clinical medicine. However, caution is still required because of the inherent risks of misinterpretation in a fully automated clinical setting. In contrast to other areas of AI technology where usage without supervision is common, AI for health should be closely monitored. Before deploying AI technology-based solutions for clinical or public use, the legal responsibility for their use must be determined. The type of harm produced by AI technology breakdown determines who bears accountability. The creator, developer, or manufacturer might be held accountable if functional flaws account for the majority of the malfunction. Before implementing AI technology, responsibilities should be well understood and assigned.

A workable plan should be in place to determine who is legally liable and what responsibility each stakeholder has in the damage if AI technology causes any harm. To limit harm, everyone involved from conception to implementation needs to work together.

Nondiscrimination and fairness principles

Another ethical issue with AI-powered diagnostics is the likelihood of bias.

The objectivity of AI systems depends on the quality of the data used for training. If the dataset is not varied and representative, biases can enter into the algorithm, resulting in erroneous and potentially dangerous diagnoses. To reduce biases, it is critical to train AI systems on diverse datasets that include patients from varied racial, ethnic, and socioeconomic backgrounds.

The following rules should be adhered to prevent biases and mistakes in algorithms and preserve quality.

1. The training algorithm's dataset neds to be precise and accurately reflect the population it will be applied.
2. In order to avoid biases and mistakes that could lead to malfunctioned AI systems, constant customer feedback, analysis, and external independent algorithmic audits should be conducted. Developers and researchers of AI must recognize any biases and appropriate steps must be taken to correct them.
3. Individual equality should be promoted and protected as a top priority for developers. Equality and justice must be applied to freedom, rights, and dignity.
4. AI technologies should be built for broad application. It is unacceptable to discriminate against people or groups based on their social status, caste, age, race, or religion.
5. If any patient/participant has been harmed, the reversibility of AI technology's decisions should be evaluated. The ability to reverse decisions must be added to the architecture of AI prior to the technology use.

6. If any unfortunate occurrences occur as a result of AI technology failing, an adequate redressal process for the victim should be in place. The maker must ensure that suitable grievance redressal mechanisms are in place.
7. There must be a secure system for raising concerns about AI technology, which can be technical, functional, ethical, or technological misuse. A proper process for protecting whistleblowers should be in place.

Validity

Before being used on patients/participants, in-depth clinical and field validation is required for AI in healthcare technology. These are required for safety and efficacy. Differences in the datasets used to train AI algorithms may amplify the divergence of AI-based systems. Such discrepancies in the diagnostic abilities of various systems are likely to confound people using them, including medical professionals and patients receiving AI healthcare, just like any other diagnostic tool. An internal procedure ought to be established to keep these issues and give developers pertinent input while minimizing the clinical setting. It is imperative to have an efficient process for feedback in place to implement necessary modifications whenever AI technology affects a person or a system of healthcare. The use of AI-based choices in clinical settings can result in clinical mismanagement or a serious health concern.

Trustworthiness

As far as AI healthcare is concerned, reliability is the most desired quality in a diagnostic or prognostic tool. Physicians must develop trust in the tools they utilize, and the same is true for AI technology. To effectively deploy AI, physicians and healthcare professionals must have a straightforward, methodical, and trustworthy method for testing the validity and dependability of AI technology.

Data quality enhancement

AI is a data-driven technology, and the outcomes are heavily reliant on the data used to train and test AI. A skewed and inadequately, large dataset may lead to issues such as discrimination, bias, and other issues, which is why this is particularly crucial in field of AI for health.

The biggest obstacle to data-driven technologies such as AI for healthcare is believed to be bias in the data. It is imperative to apply precautions to ensure that the "training data" appropriately represents significant segments of the population being studied and is free from recognized biases.

Equity, inclusivity, and accessibility

Utilizing computers for development and implementing artificial intelligence in healthcare, requires more infrastructure to be available. AI developers and relevant authorities must ensure that AI technology is distributed fairly. Special care must be given to the underprivileged and those lacking the infrastructure to use such technology. Such groups must be prioritized. Regulatory bodies such as government must promote the adoption of AI technologies by these prospective end-user groups.

Certain systems based on AI could require power, other infrastructure, internet access and technical know-how. People using AI-based technology should ensure adequate infrastructure is provided for the smooth and optimal operation of an AI technology designed for use in a low-resource scenario.

For different apps to function flawlessly across many platforms, interoperability of AI based software needs to be taken into consideration wherever possible. It supports an expanded set of accessibility options for user groups.

Collaboration

AI for health is a data-driven field. Any significant use of AI for health requires a big collection of carefully verified datasets. This can only be accomplished by promoting teamwork at all levels. Undoubtedly, enhancing the output of this field of study is likely to require a partnership among AI researchers and healthcare practitioners during the creation and deployment of healthcare-based solutions.

As a result, stronger cooperation must be established by the following principles:

While interdisciplinary collaboration is welcomed, it is critical to guarantee that no harm is done to the patients whose data may be utilized to design or test algorithms during AI technology trials.

Motivation for cooperation, as well as possible conflicts of interest, should be publicly acknowledged and thoroughly investigated if necessary to avoid harm to any stakeholders.

The Health Ministry's Screening Committee (HMSC) must give its approval before any foreign cooperation or funding related to biomedical and health research may begin. This includes data collection, sharing biological samples, and intellectual property.

In the case of incorporating healthcare data in the research and development of an AI system, data sharing is essential for any national or international collaboration while safeguarding privacy and security, as it may contain very sensitive information about a participant.

Challenges of AI in healthcare

Few genuine AI-enabled solutions being implemented in clinical practice, despite their great promise. In addition to privacy concerns, AI technology has significant methodological and technical shortcomings. The following are the principal challenges facing AI in healthcare.

Limited data

Healthcare businesses have lately utilized artificial intelligence (AI) to improve patient care and outcomes. But this is seriously hampered by a lack of adequate data. Health-related information is frequently divided and hard to obtain, making it tough to train AI models. Furthermore, many healthcare facilities need more infrastructure and resources to complete AI technology deployment. The use of artificial intelligence in health care is still lacking despite the potential advantages. Nevertheless, with continued investment and improvements in data collection and management, healthcare in India and other developing nations might change as a result of artificial intelligence.

Data bias

If the training data does not fairly represent the population that the AI system will eventually use the system may become biased. Results may be incorrect or unfair, particularly for underprivileged communities. The performance metrics for models based on AI might not necessarily be transferable to healthcare settings. The difference between the technical accuracy of AI tests and the medical efficacy that has been demonstrated in the real world is known as the AI gap. To bridge the gap, scientists and medical professionals should work together to examine how AI algorithms improve care for patients. To assess the accuracy of AI models, they can employ decision curve analysis. They can evaluate a prediction model's utility by comparing the available data sets and determining the probability that it will be successful in practice.

Lack of transparency

Because of the difficulty in understanding how an AI system arrives at a certain decision, many of them are referred to as "black boxes". Because of the inadequate transparency, it may be challenging for practitioners and other healthcare providers to accept the results of an AI system.

Lack of standardization

One of the biggest barriers to the use of AI in healthcare is the lack of uniformity. For the use of artificial intelligence in healthcare settings, there are currently no agreed standards. This lack of uniformity might cause difficulties for healthcare practitioners as well as patients.

Knowing which AI applications are efficient and which are not might be difficult for healthcare providers. This can lead to the loss of resources and time on applications based on AI that don't turn out as planned. A lack of standardization might leave people confused and unclear about the level of treatment they will get. Patients can be hesitant to adopt AI applications if they have any doubts regarding their efficacy. The lack of standardization is ultimately a major issue that must be resolved for AI to be successfully implemented in healthcare.

The use of AI in healthcare still lacks clear guidelines and standards. Due to this, it may be difficult for medical organizations to comprehend how to use technology ethically and patients may find it challenging to know what to anticipate when dealing with an AI system.

Lack of understanding

It's probable that many individuals, including healthcare experts, are unclear about how AI works and exactly what it can and can't do. This could lead to unrealistic expectations and a loss of faith in technology on the part of users. The lack of qualified employees in the AI industry is one of the most challenging issues in healthcare. Even though AI has enormous potential to revolutionize healthcare, few people are qualified to create and use AI-based technology.

The shortage of skilled people severely hinders the implementation of AI in the healthcare industry. If there aren't enough highly qualified individuals to develop and implement AI solutions, the potential benefits of AI won't be achieved. Healthcare organizations need to engage in education and training programs to help their staff members obtain the appropriate skills to address this important shortfall. Healthcare can only fully utilize the power of AI by developing a solid pipeline of talent.

Security and data privacy

For the use of artificial intelligence (AI) in healthcare, many patient records are required, which raises concerns regarding their confidentiality and safety. Providing patients with control over their data's usage and ensuring that their information is protected from unwanted access is vital.

Fear of change and lack of trust in AI

The fact that many individuals are afraid of change and have little faith in AI is one of the obstacles to the adoption of AI in healthcare. This is true when it comes to selecting the course of treatment. Even when something has the potential to make our lives better, we are frequently hesitant to give it a try.

We may even be more resistant to change when it comes to something as personal as our health. Making sure there is an honest and open discourse about how AI is being utilized in healthcare and the advantages it can provide is one method to address this difficulty. It's also crucial to establish trust by including patients and medical professionals in the creation and application of AI-based solutions. We can fully realize the potential of AI in healthcare by addressing these issues.

High cost

While there are many potential uses of AI in the healthcare industry, acceptance of AI is still hampered by the high costs of its deployment and research. To build an AI system that can accurately identify illnesses, a large amount of resources must be allocated for data collection, annotation, and algorithm development. A large infrastructure investment is also needed to implement such a system in a hospital or clinic. Because of this, the high expenses of adopting AI may continue to restrict its wider usage in healthcare.

The development of AI is being hampered by these difficulties. But happily, professionals are figuring out a way around it and addressing the issues to catalyze the development of the technology and support healthcare in more effectively easing and healing people.

Future of AI in health care

Artificial Intelligence (AI) is developing quickly and becoming more complex daily. It's only logical to question what the future of AI holds for the healthcare sector given the quick developments of AI capabilities. There are numerous possible applications for AI in the healthcare sector, from patient care and safety to diagnostic and treatment planning applications in healthcare will advance as the technology does.

Diagnostic is one field that is particularly ready for AI advancement. In order to find patterns and connections that could lead to an earlier and more precise diagnosis of diseases, using AI, huge quantities of data can be evaluated to provide new therapeutic choices depending on the unique physiology of each patient. Patient care and safety are two more areas where AI has a significant impact. In some hospitals, AI-powdered robots are already in use to help with tasks such as lifting and moving patients. Future patient care may be more heavily influenced by artificial intelligence (AI), for example, via real-time monitoring of vital signs and early detection of potentially dangerous circumstances.

AI helps medical professionals identify illnesses such as joint problems, heart problems, and cancer predictions, among others, greatly. Radiologists may now instantaneously analyze images, doing away with the need for many scans or a drawn-out review process to identify image anomalies.

Hospitals all across the world are using AI to enhance the management of resources along with imaging and diagnoses. By incorporating predictive analytic tools such as BlueDot and HealthMap with information technology departments, hospitals may utilize machine learning to effectively monitor outbreaks of infectious disease. Healthcare organizations can obtain the tools and network of support they require to stay ahead of an epidemic.

Healthcare supply chains can improve forecasting and service levels by integrating diagnostic data with logistics management. The current healthcare pandemic has drawn attention to the value of the supply chain in preserving effective healthcare systems, and the employment of AI systems increases the importance of supply chains.

Conclusion

Healthcare solutions in the future will rely heavily on AI. Because of its distinctive characteristics, precision medicine has developed into a much-needed advancement in healthcare. Despite gaining a lot of media coverage, the global healthcare business has not been significantly impacted by AI-powered solutions. They have made some little progress towards tackling important concerns. If these concerns can be handled in the coming years, it might have a significant effect on how healthcare organizations function in coming years, boosting clinical resources and guaranteeing the best patient outcome. Hence to provide patients with advanced healthcare facilities and excellent service we studied several tools and techniques. We looked at the different ethical considerations and challenges currently healthcare system has to deal with.

Despite its initial challenges, AI will eventually become efficient in both the diagnosis and treatment appears possible that eventually, a computer will review a great deal of pathology and radiology images. Speech and text recognition will be more frequently used in contexts including patient interaction and clinical note-taking.We must implement AI in the most ethical and legal manner possible, while also ensuring proper protection.

References

Alabdulatif, A., Khalil, I., Saidur Rahman, M., 2022. Security of blockchain and AI-empowered smart healthcare: application-based analysis. Appl. Sci. 12 (21). https://doi.org/10.3390/app122111039.

Almalawi, A., Khan, A.I., Alsolami, F., Abushark, Y.B., Alfakeeh, A.S., 2023. Managing security of healthcare data for a modern healthcare system. Sensors 23 (7). https://doi.org/10.3390/s23073612.

Glover, W.J., Li, Z., Pachamanova, D., 2022. The AI-Enhanced Future of Health Care Administrative Task Management. https://doi.org/10.1056/CAT.21.0355.

Hazarika, I., 2020. Artificial intelligence: opportunities and implications for the health workforce. International Health 12 (4), 241–245. https://doi.org/10.1093/inthealth/ihaa007.

Jain, K., Prajapati, V., 2021. NLP/Deep learning techniques in healthcare for decision making. Prim. Health Care Open Access 11, 1–4.

Kerasidou, C.(X.), Malone, M., Daly, A., Tava, F., 2023. Machine learning models, trusted research environments and UK health data: ensuring a safe and beneficial future for AI development in healthcare. J. Med. Ethics 49 (12), 838–843. https://doi.org/10.1136/jme-2022-108696.

Khalid, N., Qayyum, A., Bilal, M., Al-Fuqaha, A., Qadir, J., 2023. Privacy-preserving artificial intelligence in healthcare: techniques and applications. Comput. Biol. Med. 158. https://doi.org/10.1016/j.compbiomed.2023.106848.

Khurana, R., Choudhary, M., Singh, A., Singh, K.K., 2023a. AIML-based blockchain solutions for IoMT. Blockchain and Deep Learning for Smart Healthcare. Wiley, pp. 74–89.

Khurana, R., Choudhary, M., Singh, A., Singh, K.K., 2023b. Energy-efficient fog-assisted system for monitoring diabetic patients with cardiovascular disease. Blockchain and Deep Learning for Smart Healthcare. Wiley, pp. 323–352.

Kumar, Y., Koul, A., Singla, R., Ijaz, M.F., 2023. Artificial intelligence in disease diagnosis: a systematic literature review, synthesizing framework and future research agenda. J. Ambient Intell. Hum. Comput. 14 (7), 8459–8486. https://doi.org/10.1007/s12652-021-03612-z.

Kuwaiti, A.A., 2023. A review of the role of artificial intelligence in healthcare. J. Pers. Med. 13 (6), 951.

Liu, S., Liu, S., Cai, W., Pujol, S., Kikinis, R., Feng, D., 2014. Early diagnosis of Alzheimer's disease with deep learning. In: 2014 IEEE 11th International Symposium on Biomedical Imaging, ISBI 2014. Institute of Electrical and Electronics Engineers Inc., Australia, pp. 1015–1018. https://doi.org/10.1109/isbi.2014.6868045.

McCauley, N., Ala, M., 1992. The use of expert systems in the healthcare industry. Inf. Manag. 22 (4), 227–235. https://doi.org/10.1016/0378-7206(92)90025-B.

Mondal, S.S., Mandal, N., Singh, K.K., Singh, A., Izonin, I., 2023. EDLDR: an ensemble deep learning technique for detection and classification of diabetic Retinopathy. Diagnostics 13 (1). https://doi.org/10.3390/diagnostics13010124.

Morgan, A.A., Abdi, J., Syed, M.A.Q., Kohen, G.El, Barlow, P., Vizcaychipi, M.P., 2022. Robots in healthcare: a scoping review. Curr. Robot. Rep. 3 (4), 271–280. https://doi.org/10.1007/s43154-022-00095-4.

Naik, N., Hameed, B.M.Z., Shetty, D.K., Swain, D., Shah, M., Paul, R., Aggarwal, K., Brahim, S., Patil, V., Smriti, K., Shetty, S., Rai, B.P., Chlosta, P., Somani, B.K., 2022. Legal and ethical consideration in artificial intelligence in healthcare: who takes responsibility? Front. Surg. 9. https://doi.org/10.3389/fsurg.2022.862322.

Nair, I.S., et al., 2019. Natural language processing in medicine: a review. Int. J. Eng. Res. Technol. 7 (5).

Rahman, M.M., Khatun, F., Uzzaman, A., Sami, S.I., Bhuiyan, M.A.A., Kiong, T.S., 2021. A comprehensive study of artificial intelligence and machine learning approaches in confronting the Coronavirus (COVID-19) pandemic. Int. J. Health Serv. 51 (4), 446–461. https://doi.org/10.1177/00207314211017469.

Rakhra, A., Gupta, R., Singh, A., 2021. Medical Imaging and Diagnosis Using Machine Learning and Deep Learning. Informa UK Limited, pp. 390–394. https://doi.org/10.1201/9781003193838-71.

Saravanan, V., Aneja, I., Yang, H., Pillai, A.S., Singh, A., 2021. Impact of big data in healthcare system-a quick look into electronic health record systems. In: Machine Learning and the Internet of Medical Things in Healthcare. Elsevier, United States, pp. 251–262. https://doi.org/10.1016/B978-0-12-821229-5.00009-4.

Shinde, R., Patil, S., Kotecha, K., Potdar, V., Selvachandran, G., Abraham, A., 2022. Securing AI-based healthcare systems using blockchain technology: a state-of-the-art systematic literature review and future research directions. arXiv. https://doi.org/10.48550/arXiv.2206.04793.

Singh, K., Singh, A., Lin, J., 2021. Deep Learning and IoT in Healthcare systems:Paradigms and.

Sunarti, S., Fadzlul Rahman, F., Naufal, M., Risky, M., Febriyanto, K., Masnina, R., 2021. Artificial intelligence in healthcare: opportunities and risk for future. Gac. Sanit. 35, S67–S70. https://doi.org/10.1016/j.gaceta.2020.12.019.

Wu, S., Wang, J., Guo, Q., Lan, H., Zhang, J., Wang, L., Janne, E., Luo, X., Wang, Q., Song, Y., Mathew, J.L., Xun, Y., Yang, N., Lee, M.S., Chen, Y., 2022. Application of artificial intelligence in clinical diagnosis and treatment: an overview of systematic reviews. Intell. Med. 2 (2), 88–96. https://doi.org/10.1016/j.imed.2021.12.001.

CHAPTER 7

Healthcare revolution: Advances in AI-driven medical imaging and diagnosis

Amrit Suman[1], Preetam Suman[2], Sasmita Padhy[2], Naween Kumar[3] and Akansha Singh[3]
[1]Sharda School of Engineering & Technology, Sharda University, Greater Noida, Uttar Pradesh, India; [2]School of Computing Science and Engineering, VIT Bhopal University, Bhopal, Madhya Pradesh, India; [3]School of CSET, Bennett University, Greater Noida, Uttar Pradesh, India

Introduction

Big data and machine learning have had a significant impact on various aspects of modern life, including entertainment, commerce, and healthcare. AI and machine learning in these areas have demonstrated significant capabilities. We are taking a closer look at how these technologies are being utilized:

- Companies such as Netflix to analyze users' viewing habits and preferences, enabling them to provide personalized recommendations for movies and TV series, use machine learning. This enhances the user's experience and maintains viewer engagement.
- Amazon's recommendation system is based on machine learning algorithms that analyze user data to predict their preferences and provide personalized product recommendations. This is responsible for driving sales and customer satisfaction.
- The potential of AI and machine learning in healthcare is immense. They are utilized in various ways, from diagnosis to treatment.
- The analysis of medical images, including X-rays, MRIs, and CT scans, has been remarkable by AI algorithms in Medical Imaging. Their help can be used by radiologists to identify anomalies and potential health issues, which can result in faster and more accurate diagnoses.

Responsible and Explainable Artificial Intelligence in Healthcare
ISBN 978-0-443-24788-0
https://doi.org/10.1016/B978-0-443-24788-0.00007-8

- The extraction of valuable insights from electronic medical records (EMRs) can be achieved through machine learning. The integration of symptoms, biomarkers, and patient data can help in disease characterization, prognosis, and personalized treatment recommendations.

The healthcare industry faces challenges such as improving diagnostic accuracy, reducing medical errors, and enhancing patient care, which can be addressed by these applications. The enthusiasm surrounding AI's potential to transform healthcare is valid because it offers the chance to improve patient outcomes, increase efficiency, and decrease costs. The industry is set to be revolutionized by AI-driven healthcare solutions, which will lead to more proactive health management and personalized treatments. The advancement of AI in healthcare necessitates the implementation of rigorous testing and validation. Ethical considerations are necessary to maximize the benefits while minimizing potential risks (Miller and Brown, 2018). The healthcare industry is facing many challenges, such as the increased demand for services, the shortage of healthcare professionals, and rising patient expectations driven by technology giants. Innovative solutions to address these challenges are being provided by the integration of technology and artificial intelligence (AI) in healthcare.

These are some important things to keep in mind:

- Telehealth and Remote Healthcare: The use of wireless communication and smartphones to facilitate telehealth technology has become a critical component of healthcare delivery. Remote interaction between patients and healthcare providers is possible, which allows for access to healthcare services from virtually anywhere. Specialized medical care is lacking in underserved regions and locations, making this particularly valuable. Cost-effective solutions and reducing the risk of disease transmission in clinical settings can be achieved through the use of telehealth.
- Telehealth has great potential in developing countries that are expanding their healthcare systems and is relevant globally. Designing healthcare infrastructure that aligns with current needs is possible due to the opportunity to overcome geographical and resource limitations. The reliability of these solutions requires rigorous validation of their safety and effectiveness.
- Healthcare operations and service delivery will be revolutionized by AI-powered healthcare. The healthcare system can be optimized by it.
- Significant cost savings can be achieved in the healthcare sector through AI applications, with projections suggesting savings of billions of dollars. The shift from reactive healthcare models to proactive ones, which

emphasize health management and early intervention, is largely responsible for this change, which reduces the need for hospitalizations and intensive treatments.

- Continuous monitoring and personalized coaching are supported by AI technologies, which promote individual health. Aiding in early diagnoses, tailoring treatments based on patient data, and they can do facilitating more efficient follow-up care.

The utilization of AI in healthcare has the potential to improve patient outcomes (Padhy et al., 2022), reduce costs, and offer more accessible and convenient healthcare services. The safety and effectiveness of AI-powered healthcare solutions can only be guaranteed by rigorous testing and validation. Prioritizing patient data security and privacy should be a top priority when integrating these technologies, as healthcare data is highly sensitive. The role of AI in shaping the future of healthcare will increase as the healthcare landscape evolves.

Technological advancements

Technological advancements in AI and data science have experienced a remarkable surge over the past decade. A confluence of factors, including higher computer processing speeds, the availability of extensive data repositories, and a burgeoning talent pool in AI, is what makes it characterized. The rapid evolution of AI tools and technologies, including those in healthcare, has been facilitated by these elements together. The extent of AI technology, its adoption, and its profound impact on society are set to undergo a paradigm shift thanks to this transformative wave. Deep learning (DL), a crucial development in AI, has had a profound impact on how we perceive.

AI in healthcare

The healthcare industry is in agreement that AI tools are meant to complement and enhance the work of human healthcare professionals instead of completely replacing them. Healthcare personnel will be assisted by AI in a variety of tasks, such as administrative ones, clinical documentation, patient engagement, and specific functions such as medical image analysis, medical device automation, and patient monitoring (Kalis et al., 2018). AI can have a significant impact in several key areas, as identified by various reports and experts.

- Administrative workflows can be streamlined by AI, including appointment scheduling, billing, and managing electronic health records (EHRs). Healthcare professionals are able to focus more on patient care thanks to this.

- The ability of AI to interpret medical images, such as X-rays, MRIs, and CT scans, is highly proficient. Radiologists and other specialists can benefit from it in identifying abnormalities and making faster and more accurate diagnoses.
- To enhance precision and minimize invasiveness, robotic surgery is increasingly being utilized by AI-powered robots. More intricate and minimally invasive procedures can be performed by surgeons by controlling robotic systems.
- Virtual assistants, driven by AI, can help with tasks such as answering patient queries, providing medication reminders, and offering general healthcare information, thereby improving patient engagement and adherence to treatment plans.
- AI can assist clinicians in making more informed decisions about diagnosis and treatment by analyzing patient data, including medical histories and test results. Improving outcomes and patient safety can be achieved through this.
- Wearable health trackers and smart home technologies can be developed using AI to create connected and cognitive devices that keeps track of patients and gives valuable information to healthcare providers.
- The use of AI in personalized medicine enables the tailoring of treatment plans for individual patients, resulting in more effective therapies with fewer side effects.
- The precision and dexterity of surgeons can be enhanced by AI-powered robots, especially in minimally invasive procedures.
- The combination of electronics and pharmaceuticals is being used by electroceuticals to develop new therapeutic interventions, which is an emerging field. AI has the potential to help design and optimize these treatments.

The healthcare value chain is encompassed by these applications that go beyond direct healthcare delivery. Among other things, drug development (Carlton, 2020), remote patient monitoring, and ambient assisted living (AAL) for elderly or disabled individuals (Konieczny and Roterman, 2019) are included. The use of AI in healthcare is still evolving, and each of these areas has the potential to enhance patient care, improve efficiency, and drive innovation within the healthcare industry.

Artificial intelligence and medical visualization

The interpretation of medical data can be revolutionized by AI, which has the ability to recognize images, analyze patterns, and use deep learning. By assisting in the rapid and accurate analysis of medical images and videos, it can reduce the burden on human experts, make medical diagnoses, and research more efficient. AI-based systems are capable of learning from extensive datasets and adapting to new information and research, making them a dynamic and continually improving resource for medical professionals. AI's incorporation into healthcare and medical imaging has the potential to improve diagnostic accuracy and decrease analysis time. At the end of the day, dealing with the growing demand for medical expertise in the interpretation of visual data. The potential of this technological advancement lies in enhancing healthcare services and contributing to better patient outcomes.

CV for surgery and diagnosis

The field of computer vision (CV) is focused on enabling machines to interpret images and videos at a level comparable with or exceeding human capabilities. Tasks such as recognizing objects and scenes are covered in this. Various domains have benefited from computer vision, with significant contributions in fields such as image-based diagnostics and surgery guidance. Statistical signal processing was a major factor in computer vision in the past. The use of deep learning (DL) techniques is increasing in the engineering of computer vision algorithms, particularly in applications such as skin lesions and other tissue-based images. Real-time video analysis during laparoscopic procedures has been able to achieve an impressive 92.8% accuracy in identifying procedures steps and detecting missing or unexpected ones. Artificial intelligence (AI) and computer vision can be used to enhance specific surgical skills, such as suturing and knot tying, in the field of surgical technology in a remarkable way. Johns Hopkins University's smart tissue autonomous robot (STAR) has shown that it can perform better than human surgeons in certain surgical procedures (Petscharnig et al., 2018). To ensure clinical relevance, quality, and a smooth transition from the laboratory to the clinical setting, it is crucial for surgeons to be actively involved in developing these tools (Lundervold and Lundervold, 2019; Saravanan et al., 2021).

Healthcare using AR and VR

There is a significant impact on every facet of the healthcare system when Augmented Reality (AR) and Virtual Reality (VR) are integrated. From the beginning of medical education to advanced training for specialized practitioners, including experienced surgeons. It's important to acknowledge that these technologies, although beneficial, may also pose risks for patients.

Patient experience

Immersive technologies, such as VR, have shown promise in providing various benefits for individuals with debilitating conditions and those experiencing pain and discomfort (Mondal et al., 2022). In different healthcare contexts, VR can offer potential advantages, as highlighted in the passage you provided.

VR can be utilized as a rehabilitation tool for patients with chronic strokes. Patients can engage in upper limb movement exercises by engaging in immersive experiences with virtual environments and objects, such as grabbing and throwing a virtual ball. The gradual return of normal motor function can be assisted by repeated engagement, which can promote neuroplasticity.

Pain and Discomfort Management: VR offers a way to cope with pain and discomfort for patients with conditions such as cancer or mental health issues. Immersive experiences are a way to distract, relax, and entertain, which can alleviate suffering and provide much-needed respite from the challenges associated with their conditions.

Immersive technologies provide a way to escape from the real world by using artificial characters and environments. The activities of daily living can be compared with this engagement, and it can provide a valuable break from the realities of illness and discomfort.

VR is being utilized in healthcare as a growing field (Dias and Cunha, 2018), and researchers are constantly exploring new approaches to utilizing these technologies for therapeutic and supportive purposes. Immersive experiences can improve the quality of life for those with various health challenges by giving them an engaging and supportive environment that promotes physical and emotional well-being.

Intelligent personal health records

Modern healthcare can benefit both patients and healthcare providers by shifting from physician-oriented personal health records (PHRs) to patient-centric PHRs, which is a significant step. Patient-centric PHRs require some key points to consider.

Patient-centered PHRs empower patients by giving them the ability to control their health information. Patients are empowered to actively participate in their care, make informed decisions, and take ownership of their health through this empowerment. Their medical data can be accessed and managed, their conditions can be tracked, and their communication with healthcare providers can be more effective. Patient-centered PHRs help patients self-manage by allowing them to monitor their health conditions, set goals, and receive educational resources. By entering their symptoms, medication compliance, and lifestyle choices, they gain a better understanding and management of their conditions. Enhanced communication can be achieved by using these PHRs to improve communication between patients and healthcare providers. Sharing health records and updates securely with their care team allows providers to make informed decisions more easily. This may result in care plans that are more personalized and effective. Clinicians can reduce administrative tasks by allowing patients to manage their health information, schedule appointments, and access educational materials. To improve the overall efficiency of healthcare services, it is possible to redirect this time towards more critical patient care activities. The potential for improved health outcomes exists with patient-centric PHRs. Patients who are actively engaged with their health data and have easy access to resources have a greater chance of following treatment plans, resulting in improved disease management and, in some instances, prevention. Data Security and Privacy: Implementing patient-centric PHRs requires a strong emphasis on data security and privacy. Building trust in these systems requires ensuring the security and compliance of patient data with relevant regulations. The design of patient-centric PHRs should ensure interoperability with other healthcare systems and electronic health records (EHRs). Regardless of the healthcare facilities they visit, patients and healthcare providers can share data seamlessly through this. Patient-centered care is the result of the transition to patient-centric PHRs, where patients become active partners in their healthcare journey. Improving patient outcomes and satisfaction can be achieved through a more efficient and effective healthcare system. Patient-centric PHRs must be adopted and used successfully, which requires careful planning, technology infrastructure, and policies.

Health monitoring and wearables

Wearable health devices (WHDs) have brought about a significant shift in how individuals engage with and manage their health. The passage contains a few important points about the impact and potential benefits of WHDs:

- Healthcare professionals such as physicians have traditionally been relied upon by people to provide them with information about their own bodies and health.

- These devices are designed to monitor a variety of health metrics, including heart rate, activity levels, sleep patterns, and more.
- The flexibility of application is a key attribute of WHDs. These devices can be worn by users during various activities, including running, meditating, and even underwater activities. Continuous monitoring of health metrics in diverse situations is possible due to this adaptability.
- WHDs are designed to empower individuals through the provision of real-time health data through data analysis. The data generated by these devices can be analyzed by users to gain insights into their health and well-being. This gives people the power to take control of their own health management.
- Individual empowerment is achieved through WHDs. Users can now obtain health information without relying solely on frequent healthcare provider visits. They are able to make informed decisions about their health and lifestyle by having access to ongoing data.

Preventive healthcare can be improved by WHDs by enabling early detection of health issues, supporting wellness and fitness goals, and promoting a more proactive approach to health management (Shankar et al., 2022). Patient engagement and self-monitoring are the focus of a larger trend in healthcare that enables individuals to play a more active role in their own health and well-being.

- The statement suggests that the factors that determine a patient's health outcome are not complex and can be comprehended as a straightforward set of criteria. The interconnectivity of these criteria leads to their influence on each other and their collective impact on health.
- These criteria are heavily reliant on behavior, as highlighted here. The patient's health outcomes are heavily influenced by their actions and choices. The impact on health can be profoundly affected by behaviors such as diet, exercise, smoking, and adherence to medical recommendations.
- The statement suggests a comprehensive view of health outcomes. It recognizes that health is not only impacted by medical interventions or genetics, but also by the choices and actions of individuals.
- Interventions to improve health outcomes may involve efforts to promote healthy behaviors and lifestyle changes due to the emphasis on behavior-dependent factors. Health education, lifestyle modification, and preventive measures are just some of the areas that these interventions can cover.
- Patient-Centered Care emphasizes the importance of recognizing the patient's role in their own health. It is in agreement with the idea of

patient-centered care, which involves empowering individuals to make educated choices about their health and well-being.

The idea that health outcomes can be influenced by various factors is highlighted in this statement. Despite genetics and medical treatments, behaviors and choices are a significant factor in shaping an individual's health. This viewpoint emphasizes the importance of preventive care, health education, and behavioral modification in enhancing health outcomes. Users can be placed on a leaderboard with individuals who have similar or higher average weekly steps thanks to the gamification feature in a wearable device platform. The goal of this gamified scenario is to inspire users to increase their daily activities and achieve a healthier lifestyle. It acknowledges that the efficacy of these gamification features is not fully established, and there are different opinions on their benefits and potential downsides.

Below are some important points to consider:

- Competition and motivation among users are encouraged by the platform's leaderboard system. Users are encouraged to strive for higher rankings and increase physical activity by being ranked based on their average weekly steps.
- This gamification feature has a primary goal of motivating users to exercise more, which can have positive effects on their overall health. It utilizes the psychological aspects of competition and achievement to motivate behavior change.
- The lack of conclusive evidence regarding the effectiveness of gamification in wearable technology is highlighted in the passage's emphasis on effectiveness and variability. These features may not be equally beneficial to all users and their impact on user behavior and overall health outcomes may vary widely.
- Some individuals have concerns that gamification in wearables could potentially cause more harm than good, as noted. Concerns about excessive competition, overexertion, or the negative emotional impact of not ranking well on the leaderboard may be the focus of this perspective.
- The passage indicates that there is a need for more research to determine the true impact of gamification on wearable devices. This research can assist in identifying the conditions in which gamification is most effective and secure.
- Achieving a balance between the advantages and disadvantages of gamification in wearables, despite its potential to inspire healthier behavior.

Striking a balance is crucial to ensure that users are not exposed to excessive stress or risks related to excessive physical activity.

Wearable devices with gamification features aim to encourage users to lead healthier and more physically active lives. These features are still being researched and debated about their effectiveness and potential downsides. In designing and using gamification systems for health and well-being, it is crucial to balance the benefits of motivation with potential risks. Remote monitoring. Significant benefits can be offered to individuals with chronic conditions and the elderly by using smart devices and data entry.

In healthcare, remote monitoring has advantages, and here are some key points from the passage.

- Early detection of signs and symptoms of diseases or changes in health status is possible through remote monitoring. Individuals with chronic conditions are particularly benefitted by this as it provides for proactive intervention and management.
- Remote monitoring methods, which involve smart devices and manual data entry, are designed to be both convenient and minimally disruptive to people's daily lives. Patient's health status can be monitored in the background while they continue their routines.
- Through remote monitoring systems, individuals have the ability to communicate with their healthcare providers efficiently. Health data can be shared and feedback received without the need for frequent in-person visits. Healthcare becomes more convenient and accessible because of this.
- Remote monitoring's collaborative nature is highlighted by the passage between algorithms and healthcare professionals. Positive outcomes for patients are achieved through the collaboration of algorithms and smart devices with healthcare professionals. Healthcare professionals can take appropriate actions by analyzing the data and identifying trends, anomalies, or early signs of health issues using the algorithms.
- Remote monitoring is advantageous for patients because it offers a continuous and unobtrusive way to monitor their health. By taking a proactive approach, disease management can be improved, complications can be avoided, and overall quality of life can improve.
- Remote monitoring is an important aspect of modern healthcare, particularly when it comes to chronic conditions and elderly care. By allowing individuals to stay engaged in their daily activities and receive necessary healthcare support, it promotes patient-centered approach. Better health outcomes and a higher quality of care for patients can

be achieved through collaboration between technology, algorithms, and healthcare professionals.

Natural language processing

NLP is a field of artificial intelligence that concentrates on the interaction between computers and humans through natural language, such as spoken or written text.

The following are important aspects of NLP in healthcare:

- NLP aims to make computers capable of understanding, interpreting, and generating human language. Improving communication between technology and healthcare professionals or patients is crucial due to the importance of this.
- The analysis of electronic medical records (EMRs), which often contain unstructured or semistructured text, is a significant area in healthcare. NLP is beneficial in transforming this textual data into structured, machine-readable formats.
- Healthcare records contain large amounts of text data, which NLP tools are used to extract valuable information from. Patient demographics, medical histories, symptoms, diagnoses, and treatment plans are just a few of the information that can be included in this information.
- Free-text notes and narratives from healthcare providers are frequently found in EMRs when converted to unstructured data. For instance, it can automatically categorize medical records, reports, or research papers into relevant categories, which can help with information retrieval and management.
- Translation and multilingual support: NLP tools are employed to translate medical documents, reports, and patient information to facilitate communication between healthcare professionals and patients who speak different languages.

Healthcare operations become more efficient and clinical decision-making and research are supported by NLP in healthcare. By making a huge amount of textual data more accessible and understandable, public health analysis can be improved. Modern healthcare informatics is incomplete without it and it plays a crucial role in utilizing the valuable information contained in healthcare texts.

The extraction of valuable insights can be achieved through natural language processing (NLP), which plays a crucial role in healthcare. Information is obtained through unstructured textual data, such as electronic medical records (EMRs) (Petscharnig et al., 2018) and clinical notes.

Healthcare professionals find NLP particularly valuable because it enables them to interact with technology in a more natural and intuitive way.

Healthcare has a number of immediate applications for NLP:

- By extracting relevant information from physician notes and clinical narratives, NLP can streamline the billing process and make it more efficient. By using this information, accurate medical codes can be assigned, resulting in appropriate billing of healthcare services to insurance providers or patients.
- The authorization approval process is aided by NLP in extracting relevant information from physician notes and other clinical documentation. By verifying the medical necessity of procedures and treatments, this information can help prevent delays and administrative errors when obtaining approvals from insurance providers.
- NLP can be used in healthcare in a wider range of ways than just these specific applications, including but not limited to (Muragijimana et al., 2022).
- By analyzing patient records and clinical literature, NLP can assist healthcare professionals in supporting clinical decision-making and providing relevant information and recommendations.
- Healthcare organizations can improve their understanding and management of patient populations by using NLP to identify trends and insights in large volumes of patient data.
- The use of NLP is utilized for drug safety surveillance to monitor adverse drug reactions and detect potential safety issues by analyzing unstructured textual data in medical records, social media, and other sources.
- NLP allows for the exchange of patient information and medical records between healthcare providers and systems, guaranteeing that important information is accessible whenever necessary.
- The identification of eligible patients for clinical trials is assisted by NLP tools in research and clinical trials. Research studies are conducted by analyzing patient records and finding individuals with specific conditions or criteria.
- By processing information from sources such as news articles and social media, NLP is utilized in public health surveillance to monitor and analyze trends and outbreaks, supporting public health agencies in their ability to respond more effectively to emerging health threats.

The efficiency, accuracy, and effectiveness of various healthcare processes are enhanced by NLP, which is a powerful tool that is essential in modern healthcare informatics. Healthcare professionals' access and utilization of information in their daily practice could be revolutionized by its ability to interpret and extract meaning from unstructured textual data.

Clinical decision support and medical policy assessment are two of the important applications of natural language processing (NLP) in healthcare. Here are the essential details.

- NLP can be a valuable tool in supporting clinical decision-making within the healthcare team. Healthcare professionals can use data-driven insights to support their clinical decisions when using it to predict patient prognosis and outcomes.
- Compiling clinical guidance and formulating appropriate care guidelines is assisted by NLP in Medical Policy Assessment. Healthcare organizations can ensure that their policies and procedures are in line with the latest medical knowledge and best practices with the help of it.
- The ICD is a system of codes that is standardized for diseases, symptoms, and various medical findings. Clinical notes and descriptions can be utilized for NLP to categorize diseases and medical conditions.
- The accuracy of classification can be improved by using different levels of sensitivity or specificity.
- The final output is produced by aggregating the data extracted and classified by the NLP algorithm. Various healthcare applications, such as research, billing, and clinical decision support, can benefit from this structured data.
- NLP, especially when applied to disease classification and ICD coding, facilitates the transformation of unstructured clinical data into structured, machine-readable information. The accuracy and efficiency of healthcare processes can be improved by this, supporting evidence-based decision-making, and ensuring that medical practices adhere to established standards and guidelines.

Personal records integration

The creation of Electronic Medical Records (EMRs) has resulted in vast databases of patient information. Healthcare trend identification and predictive modeling can be aided by these comprehensive datasets (Dash et al., 2022a), which provide a wealth of information that can be utilized by clinicians in diagnostics and treatment decision support.

Healthcare can be revolutionized with the use of EMRs and their data, which can enable data-driven decision-making and predictive modeling. AI and data analysis tools will continue to develop, which will lead to an improvement in the ability to extract, analyze, and use this information for the benefit of patients and healthcare providers. Lastly, it leads to healthcare that is more personalized and effective. To achieve more comprehensive models of health and disease, it is crucial to close data gaps, especially during periods of wellness.

The way healthcare providers access, analyze, and utilize patient data can be transformed by AI applications in general. Healthcare professionals can gain benefits such as more accurate diagnostics, better disease predictions, and improved clinical decision support by integrating AI tools with EMRs. By making relevant data more readily available, this integration improves patient care and streamlines healthcare workflow.

Interoperability is crucial to fully leverage data in the complex landscape of Electronic Medical Record (EMR) systems. The main points are listed here.

- Different operating systems and approaches are used by different EMR providers and systems worldwide. There are more than 1000 EMR providers in the United States alone. Data integration and sharing may face challenges due to this diversity.
- The full value of healthcare data cannot be realized without the ability of EMR systems to work together and share data. Patient information can be seamlessly exchanged between different EMR systems through interoperability, resulting in more comprehensive and accessible healthcare records.
- Extracting, classifying, and correlating data from electronic medical records (EMRs) is greatly impacted by AI methods. Natural Language Processing (NLP), Deep Learning (DL), and neural networks are commonly employed as methods. NLP holds great value in processing unstructured clinical narratives and extracting structured information from them.
- The healthcare industry is constantly dealing with the challenges of EMR systems being diverse and the need for interoperability. Observational Health Data Sciences and Informatics (OHDSI) and other international collaborations demonstrate the potential for harmonizing healthcare data from different sources and countries. AI methods, such as NLP and deep learning, are crucial for understanding the vast amount of data in EMRs, which aids in better clinical decision support, research, and healthcare management.
- The DeepCare framework's data storage and analysis allows for disease progression modeling (Davenport et al., 2018; Wang et al., 2019).

Trends and patterns that may indicate the progression of a disease in a specific patient can be identified by it.

- DeepCare is capable of recommending interventions based on the analysis of EMR databases. Appropriate actions or treatments can be suggested to improve patient outcomes.
- Disease prognosis can be provided by DeepCare using the information from EMR data. It has the ability to determine the probable course of a disease for a patient and predict possible outcomes.

The wealth of information contained within EMRs can be leveraged to enhance disease management (Dash et al., 2021; Singh et al., 2022), intervention planning, and patient outcomes.

Smart devices

Robots are making significant progress in different healthcare fields, helping to complement and enhance the work.

- Surgical procedures are being performed using robots, particularly in minimally invasive surgery, where precision and dexterity are essential. Robots are used in laparoscopic operations to aid surgeons in performing delicate and complex tasks with increased precision.
- Rehabilitation and patient assistance involve using robots. Assistance with mobility and exercises can make it easier for patients to recover strength and mobility after injuries or surgeries.
- Robots can be incorporated into implants and prosthetic devices. Individuals who have lost limbs or require artificial assistance can significantly improve their quality of life with these advanced devices.
- Robots designed to interact with patients are being developed by some companies to enhance the connection between humans and machines in a healthcare context. Patient engagement can be improved by these robots, information can be provided, and patient needs can be met.
- AI technology is utilized in many of the robots being developed and utilized in healthcare. Classification, language recognition, image processing, and other tasks can be done using AI. The ability to understand and respond to human input, perform tasks with a high level of precision and adaptability is facilitated by this.

The integration of robots into healthcare is motivated by the objective of improving patient outcomes, increasing medical procedure efficiency, and supporting healthcare professionals in their work. The use of AI technology is crucial for these robots, as it enables them to understand and interact with humans and carry out tasks that require advanced computational and decision-making capabilities. With the advancement of technology, the role of robots in healthcare is likely to expand further, creating new opportunities for patient care and medical procedures.

Minimally invasive surgery

Significant improvements have been made in the outcomes of surgery. However, the practice of surgery still relies on low-tech procedures, mostly consisting of hand tools and instruments for cutting and sewing. The surgeon's ability to sense and differentiate between tissues and organs through touch is crucial in conventional surgery (Naeini et al., 2019), which often requires open surgery. Surgical technology is constantly evolving to reduce invasiveness and make surgery less disruptive for patients.

Minimally invasive surgery is expected to become safer and more efficient through the development of artificial tactile sensing (Naidu et al., 2017) with the aid of AI. Advanced tactile information will be available to surgeons to aid in precise procedures, ultimately leading to better patient outcomes and recovery. The direction of surgical innovation is brightened by this technological advancement.

In various healthcare areas, artificial tactile sensing has been utilized, one of which is breast cancer screening (Pranitha et al., 2022; Dash et al., 2022b). The use of artificial tactile sensing is not restricted to breast cancer screening alone. Other medical conditions, such as liver conditions, brain tissue examination, and detection of submucosal tumors, have also been evaluated using it. Medical examinations can be improved by using artificial tactile sensing, which can help healthcare professionals detect abnormalities and make more informed clinical decisions. The versatility and potential benefits of this technology in healthcare are underscored by its application in breast cancer screening and other medical assessments.

Neuroprosthetics

The idea of obtaining immortality or enhanced abilities through magical or medicinal beverages has deep historical roots and can be found in various ancient traditions. Both the Rig Veda and the Avesta, which are both ancient texts, talk about drinks that promise immortality or extraordinary abilities (David; Khurana et al., 2023). These ideas have been an integral part of human culture and mythology for quite some time. Advancements in science and technology, particularly in the field of AI, are allowing them to gradually take shape in the 21st century. The passage is summed up here:

- The desire for immortality and enhanced abilities has been a human longing throughout history. In ancient texts such as the Rig Veda and the Avesta, mythical drinks such as soma and hoama are mentioned, promising immortality and divine enlightenment to those who consume them.
- Advancements in technology, specifically in artificial intelligence (AI), are happening in the 21st century, which are leading to the realization of some of these ancient myths and desires.
- The potential of AI for human enhancement is being explored for its ability to aid and improve human functioning. While the majority of research is focused on helping individuals with preexisting conditions, there is an increasing interest in using AI to enhance the capabilities of healthy individuals.
- The landscape is changing as the use of AI for human augmentation is not yet widely used. There is a chance that we will see more practical applications in the coming years. AI-driven enhancements could be involved in a variety of aspects of human life and performance.

The idea of enhancing human capabilities and extending life is an enduring human fascination (Musk, 2019). While mythology used to be the main source of these ideas, modern science and technology, particularly AI, present new opportunities for achieving these long-held desires for human augmentation (Roberts et al., 2018).

Devices that help or enhance a person's nervous system, including both input and output capabilities, are known as neuroprosthetics. The goal of neuroprosthetics is to restore or enhance sensory, motor, or cognitive functions by directly interfacing with the nervous system. To enhance the quality of life and functional capabilities of individuals with various conditions, such as paralysis, limb loss, or neurological disorders, they can be utilized.

Multiple aspects of human functioning, including hearing, vision, cognition, sensory perception, and motor skills, can be greatly impacted by debilitating conditions. The function of these systems is as follows.

- User training involves educating an intelligent controller, which has the potential to be powered by artificial intelligence (AI). The system develops skills in interpreting the user's brain signals and comprehending their intentions during this training period.

- Error identification is a part of the training process. If the user tries to move an object to the left, but it ends up moving to the right. If the BMI-connected robotic hand moves in the opposite direction, the system will detect these errors.
- In order to register brain signals related to errors and make necessary corrections, the AI system utilizes reinforcement learning. By storing correct actions, the system gradually acquires the ability to accurately interpret the user's intentions.
- Various control policies can potentially be stored by the system through this learning process. Different sets of rules or behaviors for controlling external devices are represented by these policies, which can be personalized to meet the user's specific needs and preferences.
- Companies such as Neuralink seek to leverage advancements in material science, robotics, electronics, and neuroscience to address complex health problems, and this approach aligns with those goals.

The improvement of the lives of individuals with debilitating conditions can be achieved through the use of BMIs and AI-driven controllers. There are potential opportunities for them to regain or enhance their lost sensory, motor, or cognitive functions through new avenues.

Periprosthetics and BMIs are still in their infancy, and there is still plenty of room for exploration and development. The potential for these technologies to have significant benefits for patients with neurodegenerative diseases is huge. As these technologies continue to evolve and become more sophisticated. Neuroprocessing has great potential for individuals who may increasingly rely on it to improve their quality of life and maintain functional independence, in spite of the complexities involved in neurodegenerative conditions. The current advancements in this field can lead to transformative improvements in the lives of those who are affected by these diseases.

Ambient assisted living

The aging of the population has a significant impact on healthcare, necessitating effective solutions to support individuals living with chronic disorders (Padhy et al., 2021). Their goal is to maintain their independence and quality of life, especially as they strive to do so.

Smart home

A smart home's primary objective is to enhance the quality of life for its residents by offering advanced features and technologies that enhance convenience, security, and efficiency. Aging in Place (AAL) acknowledges the importance of smart homes in facilitating independent living for older adults.

- Various aspects of a person's health and well-being can be monitored remotely through smart homes' sensors. Vital signs, activity levels, and other health-related data are all part of this monitoring process. If necessary, healthcare professionals or family members can receive the collected information for real-time monitoring and intervention.
- Automated reminders for tasks such as medication management, appointments, and daily routines can be generated by smart homes. Older adults can maintain their independence and stay organized with these reminders.
- Alarms or alerts can be generated by the smart home in response to specific events or emergencies. As an example, it is capable of detecting falls and sending instant alerts to caregivers or emergency services.
- The behavior of their occupants can be analyzed by smart homes. Identifying changes in routine or behavior patterns can be aided by this analysis, which could indicate health issues or safety concerns. Detecting unusual or unexpected events within the home is also possible with this tool.
- Robotic assistance is a component of the smart home ecosystem, where assistive robots can offer physical support and assistance with daily tasks. Examples include retrieving objects, assisting with mobility, or even providing companionship.

Older adults can age in place with greater comfort and security when these AAL applications are integrated into a smart home environment. The support provided by these technologies is invaluable, helping individuals maintain their independence while also ensuring their well-being and safety are monitored and addressed as needed.

Smart homes, which have Internet of Things (IoT) architecture and low-cost sensors, provide valuable assistance to individuals with dementia. Enhanced safety and quality of life for dementia patients can be achieved through these technologies. Smart homes can be helpful for those with dementia in these ways.

- Sensors that are located throughout the home, including the bedroom, kitchen, and bathroom, are capable of monitoring daily activities. A sensor on the oven can detect whether the cooker has been utilized and remind the patient to turn it off after use, for example. This aids in preventing safety hazards that can arise from leaving appliances unattended.

- Rain sensors that are installed on windows can detect if a window has been left open during rainfall, which can be a safety alert. To avoid potential issues such as leaks or discomfort, the system can alert the patient or caregivers in these cases.
- Using sensors in the bathroom, activities can be monitored and hazards can be minimized. For instance, a bath sensor can aid in preventing the bathtub from overflowing, and a lamp sensor can ensure that lights are turned off when not required.
- Personalized reminders for daily routines, medication schedules, and safety precautions can be provided by smart home systems, promoting greater independence for dementia patients.
- Smart homes can incorporate security features along with safety-related sensors. To safeguard dementia patients and notify caregivers or family members of any unusual events or exits from the house, door sensors and surveillance cameras are utilized.
- By using these applications, individuals with dementia can live in a safe and supportive environment that offers peace of mind to both them and their caregivers. Safety concerns and daily tasks can be addressed and assisted with the use of technology. Dementia patients can continue to live independently with the help of smart homes, which contribute to their well-being and quality of life.

Sensors play an essential role in monitoring activities and ensuring safety in a smart home environment designed to aid individuals, particularly those with conditions such as dementia. Daily activities are collected by these sensors and sent to nearby computing devices. Various machine learning algorithms can be utilized to process the collected data. The process works like this.

- Activities, such as appliance usage, window status, and bathroom activities, are monitored by sensors located in various areas of the home. These activities receive continuous data collection from these sensors.
- A computing device within the home, such as a central hub or gateway, receives the sensor data.
- The data is processed by the computing device either locally or uploaded to the cloud for more advanced processing. The use of more advanced machine learning algorithms is enabled by cloud-based processing.
- Machine learning algorithms are employed to examine data and detect patterns or abnormalities.

- Classification tasks can be done using a probabilistic method called Naive Bayes Classifier.
- A statistical model that is often employed for sequential data, such as time-series information, is known as the Hidden Markov Model.
- Support Vector Machine (SVM) is a method that discriminates and is utilized in classification and regression tasks.
- Complex pattern recognition tasks are achieved through the use of Artificial Neural Networks, a machine learning model that is inspired by the structure of the human brain.
- Analyzing the data collected over time allows the system to determine a baseline for an individual's daily activities. Abnormalities are identified as deviations from this routine.
- Alerts can be sent to relatives or healthcare professionals in case the system detects abnormal behavior or safety concerns, ensuring timely intervention when necessary.
- This strategy not only enhances safety but also promotes the well-being and independence of individuals who have conditions that necessitate extra care and monitoring. The smart home is a valuable tool in healthcare and assisted living due to its ability to adapt to the specific needs and routines of residents through machine learning algorithms.

In a typical smart home or smart assistant setup, Fig. 7.1 displays the components and their connections.

- The home's sensors are strategically placed to monitor various aspects of daily life. Sensors that are capable of being used include motion detectors, temperature sensors, door/window sensors, and health-related sensors.
- Sensors are responsible for collecting data on activities, environmental conditions, and health parameters. The home has a central hub or gateway that continuously records and transmits this data.
- A computing device called a central hub or gateway is responsible for receiving data from sensors. It is responsible for processing and analyzing data at the local level.
- Data received from the sensors is processed by the central hub through local processing. Initial data analysis can be performed, activity patterns can be defined, and deviations from routines can be detected.
- By analyzing data collected over time, activity monitoring establishes a baseline for an individual's daily activities. Monitoring daily routines and detecting abnormalities is done using this baseline.

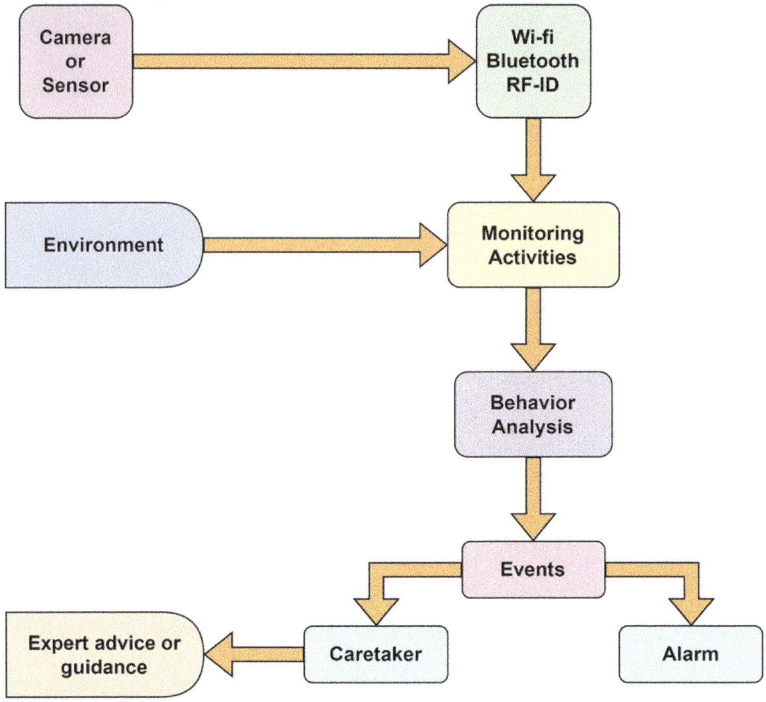

Figure 7.1 Process diagram of a typical smart home or smart assistant setup.

- Alerts and notifications can be sent by the system when it detects abnormal behavior, safety concerns, or deviations from routines. It is possible to send these alerts to family members, caregivers, or healthcare professionals.
- Through a user-friendly interface, users can interact with the smart home system. They have access to receiving alerts, checking activity logs, and customizing settings through this interface.
- The system's data collection and analysis allows for the control and automation of smart devices such as lighting, appliances, and locks in the home.
- To enhance safety, independence, and well-being for residents, the smart home system provides continuous monitoring and support.

Smart home systems designed for patients with conditions such as dementia benefit from the use of machine learning techniques, such as Markov Logic Networks. These systems have the ability to model both simple and complex activities and make decisions on appropriate alerts when patient abnormality is detected.

Assistive robots

Supporting individuals with physical limitations, particularly the elderly and those with disabilities, is crucially aided by assistive robots. These robots act as an additional pair of hands or eyes to aid in various daily tasks, enhancing the independence and well-being of their users. Assistive robots (Koumakis et al., 2019) can assist with certain activities.

- Individuals can use assistive robots to move from one place to another, such as from a bed to a wheelchair or vice versa. The importance of this support cannot be overstated for individuals who experience mobility challenges.
- Assisting with basic housekeeping tasks, such as cleaning and organizing, can help individuals maintain a tidy living environment.
- Assistive robots are capable of reminding individuals to take their medications, dispensing the necessary doses, and even monitoring medication adherence.
- Assistance during mealtime can be provided by assistive robots for those who may find it difficult to eat.
- These robots can aid in grooming tasks, such as hair brushing and shaving, and also aid in bathing for individuals.
- Social communication can be facilitated by assistive robots, which can provide companionship and conversation to combat loneliness and isolation.

RIBA (Garcia-Alonso and Fonseca, 2018) is a robot that provides assistance and has human-like arms. Various methods can be utilized to control RIBA, including tactile sensors and tactile guidance, which can be utilized to teach the robot through demonstration.

Assistive robots can greatly enhance the quality of life and promote independence for individuals with physical limitations by providing significant support and assistance. In the field of healthcare and assistive technology, they represent a valuable addition.

The MARIO project, which stands for "Managing Active and Healthy Aging with the Use of Caring Service Robots," is a project that aims to manage active and healthy aging with the aid of caring robots. The development and deployment of assistive robots is the focus of this significant initiative to address challenges. Elderly individuals are at risk of loneliness, isolation, and dementia. The goal of the project is to enhance the quality of life and well-being of the elderly population, particularly those who may be confronted with social and cognitive challenges.

The MARIO project's primary objectives could be.

- The project aims to offer companionship and social interaction to combat loneliness and isolation. The purpose of robots is to engage in conversation and activities with elderly people, serving as friendly and supportive companions.
- The project's robots can provide cognitive stimulation and memory support for individuals who are facing cognitive challenges, such as dementia, through interactive activities and reminders.
- Assistance with daily living: These robots can assist with daily activities, such as medication reminders, mealtime support, and mobility assistance, helping individuals maintain their independence.
- To guarantee the safety and well-being of elderly users, robots can have monitoring features that include fall detection and emergency response capabilities.
- The MARIO project has the potential to tailor robot interactions to the specific needs and preferences of individual users, and provide personalized care and support.
- The project is focused on ongoing research and development to enhance the capabilities and functionalities of assistive robots.

By utilizing robotics and AI technologies, the MARIO project is a promising approach to improving the lives of elderly individuals. The elderly population, especially those with cognitive challenges, can benefit from these robots' potential to provide valuable support, combat loneliness and isolation, and improve overall quality of life. The use of service robots to address the multiple needs of dementia patients has been significantly improved with the MARIO Kompa<unk> companion robot, which is a notable advancement.

The design focuses on the following objectives:

- Real feelings and emotions are conveyed by the robot, which enhances its acceptance among dementia patients. Effective care and support can be made possible by this emotional connection.
- The KompM robot aids physicians and caretakers in conducting dementia assessment tests. Tasks that involve cognitive evaluation and monitoring the patient's condition over time can be part of this.
- The robot is designed to encourage interactions with end users, promote social engagement, and provide cognitive stimulation, which are essential for the well-being of dementia patients. The Kompa<unk> robot has robotic applications that cover multiple aspects, such as:
- Users can be engaged in cognitive activities by the robot to stimulate their mental faculties.

- Social interaction aids in social interactions, providing companionship and engagement.
- Monitoring health-related aspects of users can be aided by the robot.

The MARIO Kompa<unk> companion robot, which uses advanced robotics and artificial intelligence, is meant to provide comprehensive care and support to Alzheimer's patients, while also supporting healthcare professionals and caregivers in their tasks. Technology has the potential to enhance the lives of individuals with dementia and address the challenges associated with condition.

Cognitive assistants

Those who have experienced brain injuries, such as stroke, multiple sclerosis, or trauma, are at risk for cognitive decline, which is a common issue among elderly individuals. Cognitive stimulation is a valuable approach to tackling these challenges, as it can aid in improving cognitive abilities and preventing impairment. Cognitive stimulation is increasingly being provided by assistive robots, and VIRTRAEL is one of them.

For individuals experiencing cognitive decline due to various factors, the cognitive stimulation offered by VIRTRAEL can be beneficial. Cognitive training has been demonstrated to assist in reducing cognitive impairment and improving cognitive functioning overall. VIRTRAEL enhances patients' well-being and cognitive health by customizing the program to their individual needs and offering communication and support services.

The positive impact that technology and AI can have on healthcare can be evidenced by the use of assistive robots and cognitive stimulation platforms. The increasing need for cognitive support and rehabilitation is particularly relevant in the context of aging populations.

Social and emotional stimulation

Companion robots that are designed for social and emotional interaction have shown their worth in helping elderly patients, particularly when it comes to dealing with stress, depression, and cognitive decline. Sensors and interactive features in these robots enable them to connect with patients on an emotional level and provide assistance with daily tasks.

Promising results have been demonstrated in studies on cognitive stimulation and the use of companion robots. According to their suggestion, elderly patients who interact with these robots may experience a decrease in the rate of cognitive decline and progression of dementia. The elderly's overall well-being can be positively impacted by the emotional and social support provided by these robots. Particularly those who are experiencing problems related to loneliness and cognitive impairments.

Companion robots are a part of a broader healthcare technology trend that is focused on improving the quality of life for senior citizens and addressing their unique challenges. The combination of healthcare, technology, and psychology is used by these robots to provide holistic support for elderly populations.

References

Dash, S., Panda, R., Padhy, S., 2021. Blockchain-based intelligent medical IoT healthcare system. SPAST Abstracts 1 (01).

Carlton, S., 2020. The era of exponential improvement in healthcare? Telehealth Med. Today 5 (1).

Dash, A.B., et al., 2022a. Analysis of brain function effecting form the tumour disease using the image segmentation technique. In: 2022 Second International Conference on Computer Science, Engineering and Applications (ICCSEA), pp. 1–6.

Dash, S., Padhy, S., Parija, B., Rojashree, T., Patro, K.A.K., 2022b. A simple and fast medical image encryption system using Chaos-based shifting techniques. Int. J. Inf. Secur. Priv. 16 (1), 1–24.

Davenport, T.H., Hongsermeier, T.M., Mc Cord, K.A., 2018. Using AI to improve electronic health records. Harv. Bus. Rev. 12, 1–6.

David, O., Scientific Verification of Vedic Knowledge: Archaeology Online.

Dias, D., Cunha, J.P.S., 2018. Wearable health devices—vital sign monitoring, systems and technologies. Sensors 18 (8), 2414.

Gerontechnology: first international workshop. In: Garcia-Alonso, J., Fonseca, C. (Eds.), 2018. First International Workshop on Gerotechnology. Springer.

Kalis, B., Collier, M., Fu, R., 2018. 10 promising AI applications in health care. Harv. Bus. Rev. 2–5.

Khurana, R., Choudhary, M., Singh, A., Singh, K.K., 2023. AIML-based blockchain solutions for IoMT. In: Blockchain and Deep Learning for Smart Healthcare, pp. 73–93.

Konieczny, L., Roterman, I., 2019. Personalized precision medicine. Bio. Algorithm Med. Syst. 15.

Koumakis, L., Chatzaki, C., Kazantzaki, E., Maniadi, E., Tsiknakis, M., 2019. Dementia care frameworks and assistive technologies for their implementation: a review. IEEE Rev. Biomed. Eng. 12, 4.

Lundervold, A.S., Lundervold, A., 2019. An overview of deep learning in medical imaging focusing on MRI. Z. Med. Phys. 29, 102.

Miller, D.D., Brown, E.W., 2018. Artificial intelligence in medical practice: the question to the answer? Am. J. Med. 131 (2), 129–133.

Mondal, S.S., Mandal, N., Singh, K.K., Singh, A., Izonin, I., 2022. Edldr: an ensemble deep learning technique for detection and classification of diabetic retinopathy. Diagnostics 13 (1), 124.

Muragijimana, E., Shankar, T.N., Kumar, N., Sah, B., Padhy, S., 2022. Digital crimes in cloud environment and the analysis via blockchain. In: 2022 IEEE 3rd Global Conference for Advancement in Technology (GCAT), Bangalore, India, pp. 1–6. https://doi.org/10.1109/GCAT55367.2022.9972221.

Musk, E., 2019. An integrated brain-machine interface platform with thousands of channels. J. Med. Internet Res. 21 (10), e16194.

Naeini, F.B., et al., 2019. A novel dynamic-vision-based approach for tactile sensing applications. IEEE Trans. Instrum. Meas. 1.

Naidu, A.S., Naish, M.D., Patel, R.V., 2017. A breakthrough in tumor localization: combining tactile sensing and ultrasound to improve tumor localization in robotics-assisted minimally invasive surgery. IEEE Robot. Autom. Mag. 24.

Padhy, S., Dash, S., Malla, P.P., Routray, S., Qi, Y., November 2021. An energy efficient node localization algorithm for wireless sensor network. In: 2021 IEEE 2nd International Conference on Applied Electromagnetics, Signal Processing, & Communication (AESPC). IEEE, pp. 1–5.

Padhy, S., Dash, S., Routray, S., Ahmad, S., Nazeer, J., Alam, A., 2022. IoT-based hybrid ensemble machine learning model for efficient diabetes mellitus prediction. Comput. Intell. Neurosci. 2022.

Petscharnig, S., Schöffmann, K., 2018. Learning laparoscopic video shot classification for gynecological surgery. Multimed. Tool. Appl. 77, 806.

Pranitha, G., Rukmini, T., Shankar, T.N., Sah, B., Kumar, N., Padhy, S., June 2022. Utilization of blockchain in E-voting system. In: 2022 2nd International Conference on Intelligent Technologies (CONIT). IEEE, pp. 1–5.

Roberts, A.W., Ogunwole, S.U., Blakeslee, L., Rabe, M.A., 2018. The Population 65 Years and Older in the United States: 2016.

Saravanan, V., Aneja, I., Yang, H., Pillai, A.S., Singh, A., 2021. Impact of big data in healthcare system—a quick look into electronic health record systems. In: Machine Learning and the Internet of Medical Things in Healthcare. Academic Press, pp. 251–262.

Shankar, T.N., Padhy, S., Dash, S., Teja, M.B., Yashwant, S., April 2022. Induction of secure data repository in blockchain over IPFS. In: 2022 6th International Conference on Trends in Electronics and Informatics (ICOEI). IEEE, pp. 738–743.

Singh, K.K., Sachan, V.K., Singh, A., Padmanaban, S. (Eds.), 2022. Deep Learning in Visual Computing and Signal Processing. CRC Press.

Wang, F., Casalino, L.P., Khullar, D., 2019. Deep learning in medicine—promise, progress, and challenges. JAMA Intern. Med. 179, 293–294.

Further reading

Blaschke, T., Olivecrona, M., Engkvist, O., Bajorath, J., Chen, H., 2018. Application of generative autoencoder in de novo molecular design. Mol Inform 37 (12), 1700123.

Bresnick, J., 2017. Artificial Intelligence in Healthcare Market to See 40% CAGR Surge.

Coley, C.W., Barzilay, R., Green, W.H., Jaakkola, T.S., Jensen, K.F., 2017. Convolutional embedding of attributed molecular graphs for physical property prediction. J. Chem. Inf. Model. 57 (8), 1757–1772.

Dash, S., Padhy, S., Parija, B., Rojashree, T., Patro, K.A.K., 2022c. A simple and fast medical image encryption system using Chaos-based shifting techniques. Int. J. Inf. Secur. Priv. 16 (1), 1–24. Marr B. How is AI used in healthcare—5 powerful real-world examples that show the latest advances. Forbes; 2018.

Dash, S., Padhy, S., Azad, S.M.A.K., Nayak, M., 2023a. Intelligent IoT-based healthcare system using blockchain. In: Swarnkar, T., Patnaik, S., Mitra, P., Misra, S., Mishra, M. (Eds.), Ambient Intelligence in Health Care. Smart Innovation, Systems and Technologies, vol 317. Springer, Singapore. https://doi.org/10.1007/978-981-19-6068-0_30.

Dash, A.B., Dash, S., Padhy, S., Das, R.K., Mishra, B., Paikaray, B.K., 2023b. Identification of polyp from colonoscopy images by deep belief network based polyp detector integration model. EAI Endorsed. Trans. Perv. Health Tech. 25. Available from: https://publications.eai.eu/index.php/phat/article/view/3964.

Dash, S., Padhy, S., Devi, S.A., Patro, K.A.K., 2023c. An efficient intra-inter pixel encryption scheme to secure healthcare images for an IoT environment. Expert Syst. Appl., 120622. Joseph A, Christian B, Abiodun AA, Oyawale F. A review on humanoid robotics in healthcare. In: MATEC Web of Conferences; 2018.
D'Onofrio, G., et al., 2019. MARIO Project: validation and evidence of service robots for older people with dementia. J. Alzheimers Dis. 68, 1587.
Ekins, S., et al., 2019. Exploiting machine learning for end-to-end drug discovery and development. Nat. Mater. 18 (5), 43-41.
Frendø, M., Konge, L., Cayé-Thomasen, P., Sørensen, M.S., Andersen, S.A.W., 2020. Decentralized virtual reality training of mastoidectomy improves cadaver dissection performance: a prospective, controlled cohort study. Otol. Neurotol. 41 (4).
Hashimoto, D.A., Rosman, G., Rus, D., Meireles, O.R., 2018. Artificial intelligence in surgery: promises and perils. Ann. Surg. 268, 706.
Kadurin, A., Nikolenko, S., Khrabrov, K., Aliper, A., Zhavoronkov, A., 2017. druGAN: an advanced generative adversarial autoencoder model for de novo generation of new molecules with desired molecular properties in silico. Mol. Pharm. 14 (9), 3098.
King, D., DeepMind's Health Team Joins Google Health.
Kirch, D.G., Petelle, K., 2017. Addressing the physician shortage: the peril of ignoring demography. JAMA 317 (19), 1947−1948.
Kulski, J.K., 2020. Next-generation Sequencing—An Overview of the History, Tools, and 'omic' Applications.
Lavecchia, A., 2019. Deep learning in drug discovery: opportunities, challenges and future prospects. Drug Discov. Today 24 (10), 2017−2032.
Lee, K.-F., 2019. AI Superpowers: China, Silicon Valley, and the New World Order, first ed. Houghton Mifflin Harcourt.
Lee, S.H., Jung, H.Y., Yun, S.J., Oh, B.M., Seo, H.G., 2020. Upper extremity rehabilitation using fully immersive virtual reality games with a head mount display: a feasibility study. Pharm. Manag. PM R 12, 257.
Love-Koh, J., et al., 2018. The future of precision medicine: potential impacts for health technology assessment. Pharmacoeconomics 36 (12), 1439−1451.
Mayr, A., Klambauer, G., Unterthiner, T., Hochreiter, S., 2016. DeepTox: toxicity prediction using deep learning. Front. Environ. Sci. 3, 80.
Merk, D., Friedrich, L., Grisoni, F., Schneider, G., 2018. De novo design of bioactive small molecules by artificial intelligence. Mol Inform 37.
Mondal, S.S., Mandal, N., Singh, A., Singh, K.K., 2020. Blood vessel detection from Retinal fundas images using GIFKCN classifier. Proc. Comput. Sci. 167, 2060−2069.
Padhy, S., Alowaidi, M., Dash, S., Alshehri, M., Malla, P.P., Routray, S., Alhumyani, H., 2023. AgriSecure: a fog computing-based security framework for agriculture 4.0 via blockchain. Processes 11 (3), 757. https://doi.org/10.3390/pr11030757.
Panda, R., Dash, S., Padhy, S., Das, R.K., 2023. Diabetes mellitus prediction through interactive machine learning approaches. In: Next Generation of Internet of Things. Springer, Singapore, pp. 143−152.
Rangasamy, A.S.S., Nadenichek, R., Rayasam, M., 2018. Natural Language Processing in Healthcare.
Shi, T., et al., 2019. Molecular image-based convolutional neural network for the prediction of ADMET properties. Chemometr. Intell. Lab. Syst. 194, 103853.
Singh, A., Singh, K.K., 2023. Blockchain and deep learning: research challenges, open problems, and future. In: Blockchain and Deep Learning for Smart Healthcare, pp. 425−431.
Wallach, H.A., Dzamba, M.I., 2015. AtomNet: a deep convolutional neural network for bioactivity prediction in structure-based drug discovery. arXiv.
Wu, Z., et al., 2018. MoleculeNet: a benchmark for molecular machine learning. Chem. Sci. 9.
Zhang, L., Tan, J., Han, D., Zhu, H., 2017. From machine learning to deep learning: progress in machine intelligence for rational drug discovery. Drug Discov. Today 22 (11), 1680.

CHAPTER 8

A deep learning approach for medical image classification using XAI and convolutional neural networks

Lesia Mochurad[1], Andrii Ilkiv[1] and Yulianna Mochurad[2]
[1]Lviv Polytechnic National University, Department of Artificial Intelligence, Lviv, Ukraine; [2]Danylo Halytsky Lviv National Medical University, Lviv, Ukraine

Introduction

In the modern world, where the population of our planet is constantly growing every day, the need to provide timely medical care is a priority for many medical institutions. A large number of studies on the classification of the patient's condition at the diagnostic stage indicate the relevance of this topic (Altameem et al., 2022; Chumachenko et al., 2023; Mochurad and Panto, 2023).

Distinct diseases require varied identification methods, such as X-ray scans for pneumonia, COVID-19, and lung cancer, whereas skin cancer and Alzheimer's disease can be detected through biopsies and CT scans, respectively (Goyal et al., 2022). Timely disease detection is crucial for effective treatment and preventing lasting complications. The application of deep learning has significantly impacted the medical field, enabling accurate and reliable early disease prediction (Mondal et al., 2023; Singh and Kant Singh, 2024).

These models play a vital role in supporting healthcare professionals in disease detection. However, their training heavily relies on secondary data sources, as healthcare institutions are cautious about sharing individual patient data to uphold privacy standards. This limitation affects the efficacy of deep learning models, as achieving optimal results often requires large datasets for training (Sharma and Guleria, 2023).

Responsible and Explainable Artificial Intelligence in Healthcare
ISBN 978-0-443-24788-0
https://doi.org/10.1016/B978-0-443-24788-0.00008-X

The integration of technology and healthcare has opened avenues for more accessible and precise disease diagnosis (Rakhra et al., 2021). Notably, the introduction of emerging fields in imaging science, such as radiomics and radiogenomics, has significantly influenced healthcare solutions, enhancing the effectiveness of patient condition diagnosis (Hussain et al., 2023).

While medical image segmentation models have proven valuable, their challenge lies in generalizing effectively to new datasets, limiting their applicability in clinical settings. Approaches to domain generalization, such as random convolution, which involves randomizing weights in the initial layers of CNN models, have shown promise in enhancing model generalization (Zhang et al., 2020). However, the indiscriminate introduction of high-frequency noise during early feature extraction can compromise crucial fine details and diminish model performance on unfamiliar datasets.

The challenge of medical image classification using deep learning is to accurately classify medical images into different categories based on their content, such as identifying the presence or absence of a specific disease or classifying images into different types of tissues or organs. Medical image classification is a key task for medical diagnosis and treatment, as it provides important information about the patient's condition and helps to make medical decisions (Jangra et al., 2023).

Traditionally, medical image classification has relied on manual interpretation by trained professionals, which can be time-consuming and prone to human error. However, deep learning methods have shown promising results in automating medical image classification, with several of the studies analyzed above demonstrating high levels of accuracy and efficiency.

The rationale behind the use of deep learning for medical image classification is the potential to improve the accuracy and efficiency of diagnosis and treatment. With the increasing availability of medical image data, deep learning methods can become a valuable tool for analyzing and interpreting this data, allowing medical professionals to make faster and more accurate diagnoses.

In addition, deep learning methods can help overcome some of the limitations of traditional medical image analysis methods, such as interobserver variability, limited availability of trained experts, and time-consuming analysis processes (Bo et al., 2023). By automating the classification of medical images, deep learning methods can improve the quality and speed of medical diagnosis and treatment, which can have a significant impact on diagnostic outcomes.

The main goal of this paper is to create a neural network based on deep learning using an ensemble of CNN for diagnosis classification and segmentation of computed tomography scan areas.

The practical value of this paper lies in the implementation of a user interface for diagnosing diseases with a multipurpose and the use of explanatory artificial intelligence methods.

The novelty of this paper can be summarized as follows.

- A detailed analysis of the current state of the investigated problem was conducted.
- A neural network based on deep learning, using an ensemble of CNN for diagnosis classification, was proposed.
- Model interpretability was investigated using the LIME and SHAP methods.
- The performance indicators of the model were analyzed through training on two preprocessed datasets.

The rest of the text follows this structure: In Section State-of-the-arts, a review of the literature on the topic explored in the paper is presented. Section Materials and methods describes data preprocessing, the construction of an optimal deep learning-based classification model architecture, and the training of the model. In Section Modeling and results, extensive experiments on the classification of medical images on two datasets are conducted. Section Analyzing the behavior of classification models when applying XAI methods presents the results of studying the behavior of classification models when applying XAI methods to Datasets 1 and Dataset 2. The final section includes conclusions and prospects for further research.

State-of-the-arts

Mahapatra et al. (2022) proposed a new approach to developing deep learning models for medical image analysis tasks which includes interpretation-driven inductive bias. An inductive bias is a set of assumptions or biases that a machine learning model uses to make predictions. The authors of the paper claim that interpretability should be a key component of inductive bias for medical image analysis tasks, as it can help doctors and researchers understand how the model arrives at its predictions. To achieve this, the authors propose a framework that combines deep learning with interpretable models such as decision trees and linear models. These interpretable models are used to guide the training process of a deep learning model to learn interpretable data representations. The authors apply their framework to the task of classifying pulmonary nodules on computed tomography and show that their model achieves competitive performance, and also provides interpretable insights into how the model makes its predictions.

Zhao et al. (2023) proposed a deep learning approach for localization and determination of medical image orientation. The authors of the article note that accurate localization and orientation of medical images is crucial for many medical applications, such as image registration, surgical planning, and disease diagnosis. However, this task can be challenging due to the large variability in image appearance and orientation. To solve this problem, the authors propose a deep learning solution that consists of a two-stage pipeline. In the first stage, a deep learning model is trained to localize a region of interest (ROI) in an image. ROI is a domain that contains an interesting anatomical structure. In the second stage, another deep learning model is trained to detect the orientation of the ROI. The authors demonstrate the effectiveness of their approach on two different medical image datasets: the publicly available Montgomery County X-ray (MCX) dataset and a computed tomography (CT) dataset collected independently. They show that their approach provides high accuracy in both ROI localization and orientation determination.

Jin et al. (2022) investigated a deep learning approach for recognizing objects in medical images that is guided by anatomical information. The authors of the article note that object recognition in medical images can be difficult due to the variability of image appearance and the presence of anatomical structures that can cover or partially cover the object of interest. To solve these problems, the authors propose a deep learning approach that is guided by anatomical information. Specifically, the authors use a two-stage pipeline that combines a CNN with a graph convolutional network (GCN). In the first stage, the CNN is trained to extract features from the image, and these functions are used to create a graph that represents the anatomical relationships between the various structures in the image. In the second stage, the GCN is used to spread the information over the graph to refine the results of object recognition. The authors demonstrate the effectiveness of their approach on two different medical image datasets: the publicly available Lung Image Database Consortium (LIDC-IDRI) dataset and an MRI knee image dataset. They show that their approach provides cutting-edge performance on both datasets for object recognition tasks such as nodule detection in lung CT scans and cartilage segmentation in knee MRI images.

Wang et al. (2022) proposed a deep learning approach for segmenting COVID-19 pneumonia infections in medical images using a spatial classification network and semisupervised learning. The authors of the article note that accurate segmentation of COVID-19 pneumonia infections in medical images is crucial for the diagnosis and treatment of the disease. However, obtaining a large labeled dataset to train such models is difficult because of the labor-intensive and expensive process of annotating medical images. To solve this problem, the authors propose a semisupervised multiple training case approach that uses a small labeled dataset and a larger unlabeled dataset to improve model performance. The proposed SSA-Net model consists of an encoder-decoder architecture with a spatial classification module. The self-monitoring module allows the model to focus on important image elements for accurate segmentation, while the semisupervised multiimage approach allows the model to effectively use both labeled and unlabeled data for training. The authors evaluate their proposed approach on two different COVID-19 infection segmentation datasets, including the COVID-19—20 dataset and the COVID-19-CT-1000 dataset. They show that their approach outperforms several state-of-the-art segmentation models and achieves competitive results with a small amount of labeled data.

Wang et al. (2022) proposed a BCNN architecture specifically for the task of polyp segmentation in colonoscopy images. They emphasize the difficulty of accurately segmenting polyps, which are small and often irregularly shaped and can merge with the surrounding tissue. The BCNN model includes a boundary refinement module that helps to improve the accuracy of segmentation results along the edges of polyps. The model also uses a combination of CNN and recurrent neural networks (RNN) to capture local and global contextual information in colonoscopy images. The authors evaluate the BCNN model on several colonoscopy image datasets and compare its performance with other state-of-the-art segmentation methods. They demonstrate that the BCNN model achieves higher segmentation accuracy and Dice similarity coefficient (a common measure for evaluating segmentation performance) than other methods. Overall, the paper demonstrates the effectiveness of the BCNN architecture for the challenging task of polyp segmentation in colonoscopy images and emphasizes its potential for other medical image segmentation applications.

In (Pereira et al., 2022), a new method for classifying melanoma using light fields and a combination of the Morlet scattering transform and CNNs is presented. Melanoma is a type of skin cancer that can be difficult to diagnose accurately, and early detection is crucial for successful treatment. The authors propose to use light field images that capture both the intensity and direction of light rays in a scene to improve the accuracy of melanoma classification. They also include surface depth information, which measures the distance between the skin surface and the melanoma lesion, as a valuable tool to improve melanoma detection rates. The Morlet scattering transform is used to extract features from the light field images, and the CNN is trained on these features to perform melanoma classification. The authors evaluate the performance of their method on a dataset of skin lesion images and compare it with several other cutting-edge methods. The results show that the proposed method provides high accuracy for melanoma classification, outperforming other methods on the same dataset. The authors also demonstrate the importance of including surface depth information, which significantly improves the detection rate of melanoma lesions.

In the study outlined in Liu et al. (2021), a novel two-stage cross-domain transfer learning framework is introduced for segmenting lung infections related to COVID-19 in chest CT images. The authors underscore the significance of precise segmentation in the diagnosis and treatment of COVID-19, while also acknowledging the challenges stemming from the limited availability of labeled data for this particular task. The proposed framework is structured into two stages. In the initial stage, the source domain model undergoes training on an extensive dataset of lung computed tomography data encompassing various diseases. This phase aims to capture common characteristics and enhance the model's ability to generalize to novel data. Subsequently, in the second stage, the target domain model is fine-tuned using a smaller dataset consisting of COVID-19 CT images. This process adapts the learned features to the specific task of segmenting COVID-19 infections. The authors assess the system's performance on a COVID-19 CT image dataset and compare it with several alternative segmentation methods. The results demonstrate that the proposed framework achieves high accuracy, outperforming other methods on the same dataset. Additionally, the authors highlight the effectiveness of transfer learning in enhancing segmentation performance, particularly in scenarios where there is an insufficient amount of labeled data.

Zhang and Chung (2021) suggested a new method for quantizing neural networks used in medical image segmentation to reduce memory and computation requirements while maintaining high accuracy. The authors emphasize the importance of efficient medical image analysis in clinical practice and the challenges associated with the large size of neural networks used for segmentation. This method, named MedQ, is an ultra-low-bit quantization method that can compress neural network weight coefficients to just a few bits while still preserving segmentation accuracy. The key contribution of MedQ is its lossless quantization method, meaning it introduces no additional approximation errors during the quantization process. The authors evaluate the performance of MedQ on several medical image segmentation tasks, including brain tumor segmentation and lung nodule segmentation, comparing it with other quantization methods. The results demonstrate that MedQ achieves higher accuracy than other methods using significantly fewer bits for quantization.

Bortsova et al. (2021) investigated the vulnerability of medical images analysis systems to aggressive attacks, which specially created input images cause to provoke wrong classification or other unexpected behavior in the system. The authors highlight the potential risks of such attacks in clinical practice and the need to improve the reliability of medical image analysis systems. They examine several factors that can affect the vulnerability of medical image analysis systems to adversarial attacks, including the dataset used for training the system, the type of medical image modality, and the type of attack algorithm employed. They conduct experiments with various medical image analysis tasks, such as chest X-ray classification and skin lesion segmentation, and assess the effectiveness of different models under aggressive attacks. The results show that medical image analysis systems are vulnerable to aggressive attacks, and this vulnerability depends on several factors. For example, the vulnerability is higher when the model trains on a smaller dataset or when the attack algorithm is specifically designed to exploit features of the medical image modality. The authors conclude that there is a need for further research into the vulnerability of medical image analysis systems to aggressive attacks, and that additional measures are necessary to enhance the reliability of these systems. This includes incorporating adversarial training methods during the model training process and conducting a more detailed assessment of model effectiveness under various attack scenarios.

He et al. (2021) proposed a new neural network architecture named the Multi-Attentional Feature Network (MAFNet) for assessing brain age based on magnetic resonance imaging (MRI) data. They emphasize the importance of brain age assessment in understanding the aging process and identifying early signs of neurological disorders, while addressing the challenges posed by the complex and heterogeneous nature of brain imaging data. MAFNet's architecture includes multiple channels processing different types of MRI data, such as structural, functional, and diffusion-weighted images, and a fusion module that integrates information from these channels. The authors also incorporate attention mechanisms to selectively emphasize the most informative features for age assessment. They evaluate MAFNet's effectiveness on a large dataset comprising 16,705 MRI scans of healthy individuals aged from 0 to 96 years. They compare MAFNet's performance with several other state-of-the-art methods and demonstrate that MAFNet achieves higher accuracy and resilience to variations in image quality and demographic factors. The authors also conduct an extensive interpretative analysis to explore the neural mechanisms underlying brain aging and the features most crucial for age assessment. They show that MAFNet can accurately assess brain age in various scenarios, including in healthy individuals and those with mild cognitive impairments. This paper presents a promising approach to brain age assessment using a new neural network architecture that incorporates attention mechanisms and can process multiple types of MRI data. The authors also demonstrate the potential of their method for understanding the aging process and detecting early signs of neurological disorders.

Chaitanya et al. (2021) addressed the problem of medical image segmentation, a critical task in medical image analysis. Medical image segmentation involves identifying and delineating interesting structures in medical images, such as organs, tumors, or lesions. A major challenge in medical image segmentation is the limited availability of annotated data, which hampers the training of accurate and reliable segmentation models. To overcome this challenge, the authors propose a new semisupervised data augmentation method for medical image segmentation. This method uses unlabelled data to enhance the performance of the segmentation model and reduce the reliance on labelled data. Specifically, the method consists of three steps: creating augmented images using a task-driven data augmentation method, training the segmentation model on both labelled and augmented data, and refining the segmentation model using unlabelled data. The task-driven data augmentation method used in step 1 generates augmented images tailored to the segmentation task by maximizing a segmentation-specific target function. The authors demonstrate that this method can generate high-quality augmented images that improve the performance of the segmentation model. They evaluate the proposed method on two medical image segmentation tasks: liver segmentation and lung nodule segmentation. They compare the performance of their method with several other state-of-the-art methods and demonstrate that their method achieves higher segmentation accuracy with less labelled data.

Hyun et al. (2021) focused on solving underdetermined inverse problems in medical imaging using deep learning. In many medical imaging applications, the number of measurements is less than the number of unknowns, making the problem underdetermined and leading to multiple possible solutions. The authors propose a deep learning-based approach, named Deep Learning-based Solvability Enhancement (DL-SE), which uses a deep neural network to reconstruct the underlying image from underdetermined measurements. They also include a regularization term in the loss function to promote sparsity and improve the solvability of the problem. The authors evaluate the effectiveness of DL-SE in two medical imaging applications: compressed sensing MRI and X-ray computed tomography (CT). They compare the performance of DL-SE with several other state-of-the-art methods and demonstrate that DL-SE achieves higher accuracy and faster convergence. They also conduct extensive experiments to explore the impact of various factors, such as the number of measurements, noise level, and network architecture, on the performance of DL-SE. They demonstrate that DL-SE is robust to variations in these factors and can achieve high accuracy even with a limited number of measurements.

Based on the critical analysis conducted above, the best approach for diagnosis is to implement an ensemble of CNN to work with different types of medical images.

Materials and methods

Analysis of selected datasets

The classification of the patient's disease type will be carried out using a trained ensemble of neural networks, which will be trained using two datasets. These datasets contain images obtained by computed tomography.

The Dataset 1 considered (Chest X-Ray Images (Pneumonia), 2018), intended for classifying lung conditions, consists of a collection of medical images divided into subsamples grouped by disease type (see Fig. 8.1). The dataset contains a significant number of images with a resolution of 224 × 224 pixels. It covers two main classes: "Pneumonia" and "Normal Lung Conditions". It is important to note that the class balance in the dataset may vary.

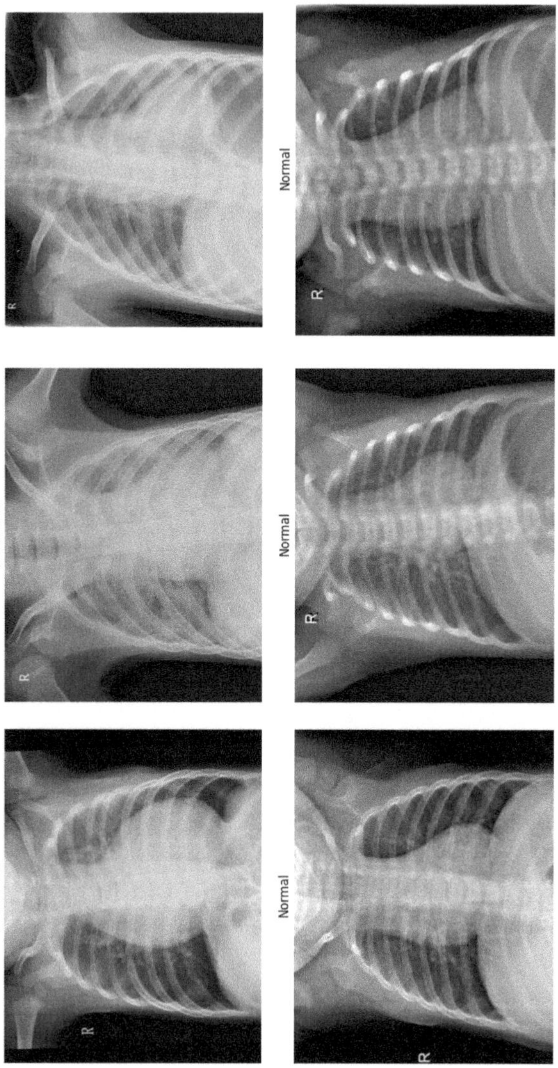

Figure 8.1 Example of input data images from the Dataset 1.

The dataset's data comes from medical archives and hospital databases, collected either with patient consent or for scientific research purposes. The images, after going through an anonymization process, meet the standards for medical data storage and confidentiality. Qualified medical professionals created annotations for each image, identifying whether or not there is pneumonia present (refer to Fig. 8.2).

This dataset is designed for developing and testing deep learning models for the automatic classification of pneumonia from lung images. It's also available for public use and download for research and educational purposes, aiming to further advance research in this important field.

The first dataset will consist of chest X-ray images and will have two classes of images.

- Chest X-ray of a healthy person
- Chest X-ray showing signs of viral pneumonia

This dataset includes 5863 X-ray scans and excludes images with potential anomalies that could be identified as probable diseases.

The Dataset 2 (Brain Tumor, 2020) features brain computed tomography (CT) images (refer to Fig. 8.3). It contains 2030 images, with a 1:1.2 ratio of images with tumors to images of healthy brains.

Each image gets classified as either "with tumor" or "without tumor" (see Fig. 8.4), setting the task to identify tumors based on MRI images. The input images vary in size, so their dimensions will be standardized to 224 by 224 pixels.

To ensure effective training of the models and improve their performance, the images underwent preprocessing. This included normalizing the images, involving subtracting the average pixel value and dividing by the standard deviation for each channel (red, green, blue). This process ensured stability and balance in the input data for the models.

Additionally, the dataset used data augmentation to expand its variety. This technique involved adding random rotation, shift, and brightness changes to the images. Augmentation helped increase data diversity and prevented the models from overfitting.

To determine the effectiveness of our models and validate their overall performance, we split the data into three main parts. The training set, containing 70% of the data, was for training the models. The validation set, comprising 15% of the data, was for tuning parameters and assessing training outcomes. The test set, also 15%, remained independent for the final evaluation of the models' performance.

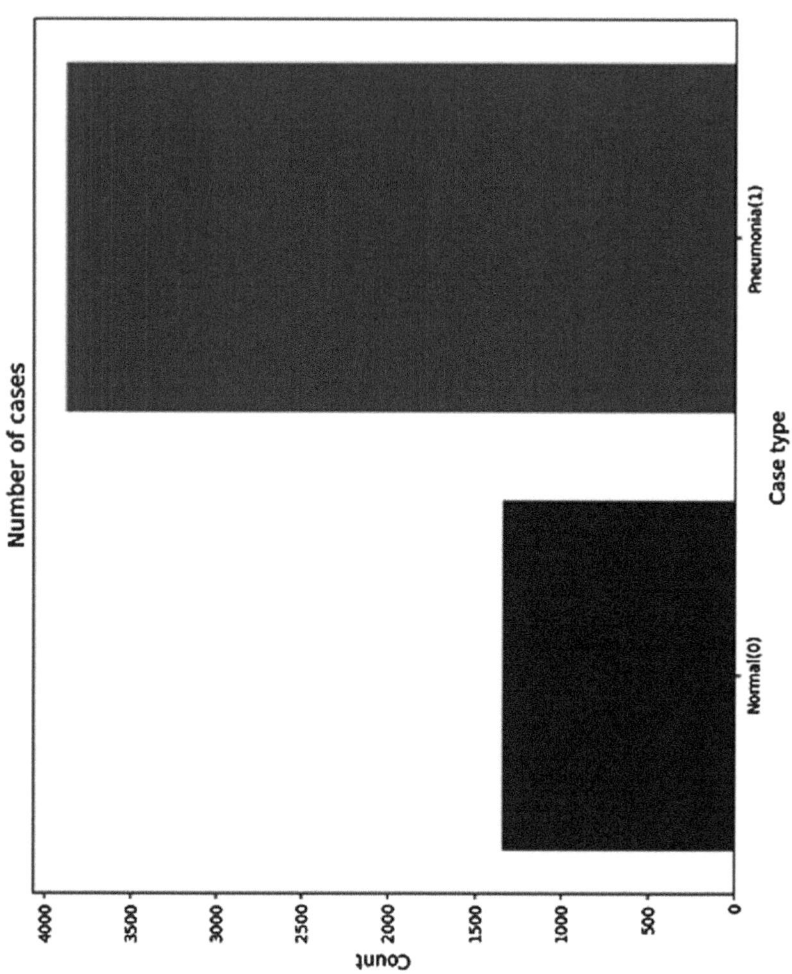

Figure 8.2 Chart comparing the class distribution for the Dataset 1.

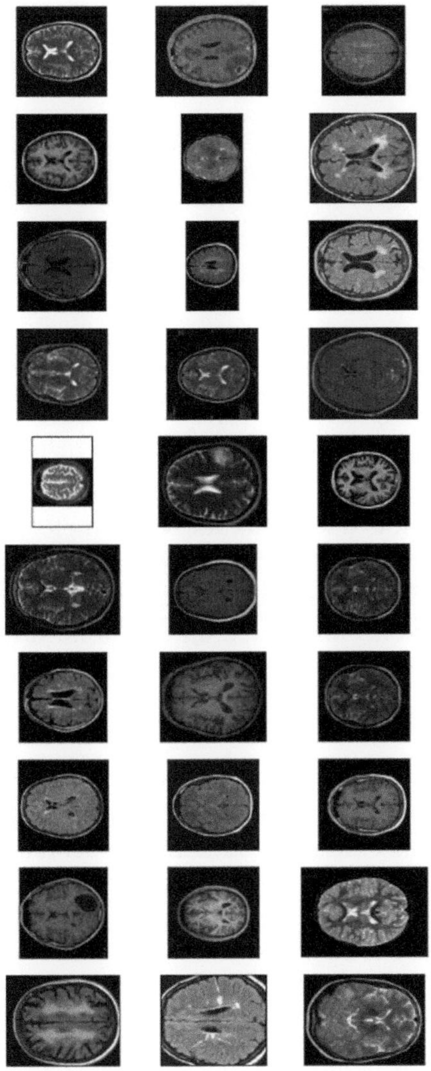

Figure 8.3 An example of the input data from the Dataset 2.

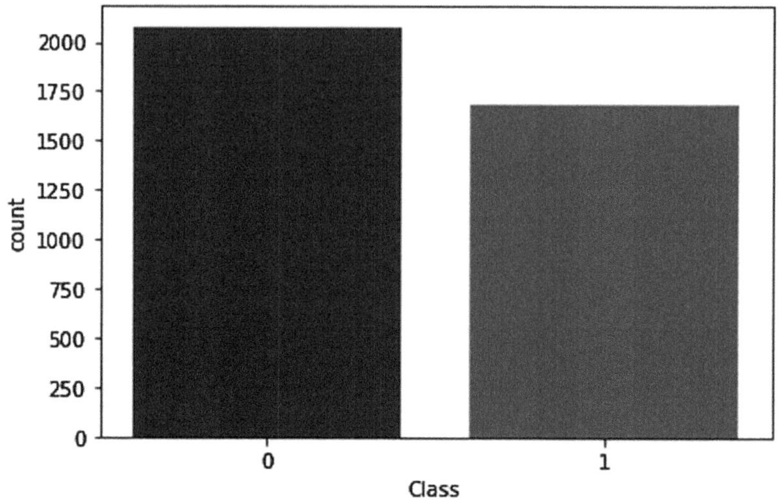

Figure 8.4 Chart comparing the class distribution for the Dataset 2.

The sample splitting parameters were chosen in such a way as to extract as much data as possible for the training process and avoid obtaining a false estimate during validation due to insufficient sample size (Paithane and Kakarwal, 2023).

Analyzing the class distribution in our data, it was determined that the classes are not always evenly distributed. In case of imbalance of classes, we either applied augmentation to smaller classes to increase the number of examples of these classes or gave weight to classes during training to account for this imbalance.

To easily and effectively utilize the data for model training and validation, it was transformed into tensor format and prepared for input into the models. This step was crucial to ensure the proper functioning of neural networks.

The proposed model architecture

Next, the architecture of the models used in our ensemble for the task of classifying the patient's condition based on diagnostic images obtained from computer tomography is proposed. Our ensemble consists of two models.

- Depthwise Convolutional Network (DBN);
- VGG-16 convolutional network with a modification.

The ensemble will be formed using the "majority voting" methodology, where each model predicts the class for the input data. Based on these predictions, the class that was chosen by the majority of models is selected. This chosen class becomes the final prediction of the ensemble. For comparison, an ensemble will also be formed using the methodology of "coordinated prediction," where each model in your ensemble provides numerical scores for each class. Using these scores, they are averaged to obtain final scores. The class with the highest average score becomes the final prediction class.

The Depthwise Convolutional Neural Network architecture was used for the convolutional neural network. Depthwise Convolution is a key component in modern deep learning architectures aimed at processing images and reducing the number of network parameters. This approach has become crucial for achieving less complex and faster models without sacrificing accuracy. Depthwise Convolution is an operation that acts on each channel (depth) of the input image individually. Typically, it is used before the Pointwise Convolution (1 × 1 convolution), which combines the output channels into a single result. This architectural feature makes Depthwise Convolution highly efficient in reducing the number of parameters and computational costs. Fig. 8.5.

As an input to this network, we receive a tensor of a certain dimension. For each channel of the input tensor, a separate convolution is performed, which is applied to each channel separately.

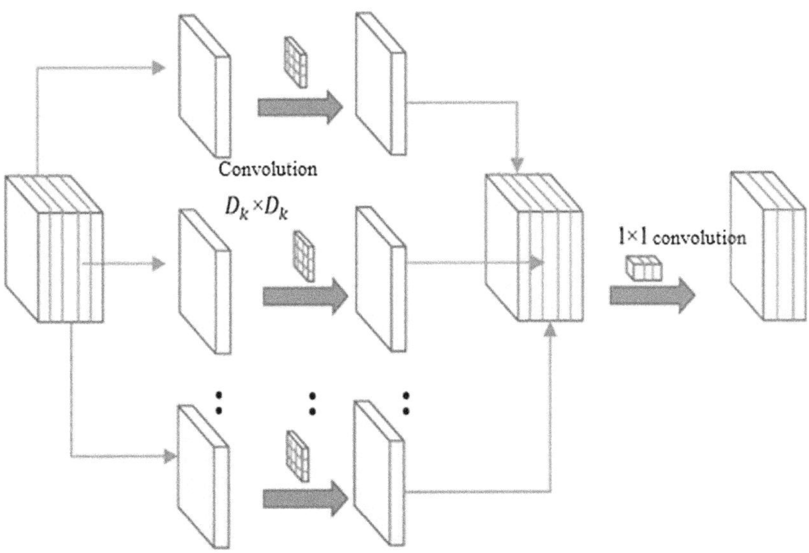

Figure 8.5 Structure of layers in the depthwise convolutional network (Sun et al., 2023).

The result is a tensor in which each channel of the output image contains information obtained from the corresponding channel of the input image.

For each increase in discretization, a function of the same size is associated, and information lost during the encoder pooling process is conveyed to the corresponding decoder at various stages through transition connections. This mechanism enhances the capability of extracting feature information (Sun et al., 2023).

The architecture of VGG16 comprises 13 convolutional and three fully connected layers, utilizing 3 × 3 kernels for the convolutional layers and 2 × 2 parameters for the pooling layers. The convolutional and pooling layers of VGG16 are organized into blocks 1—5, with each block containing several convolutional layers and one pooling layer. In block 1, each of the two convolutional layers utilizes 16 kernels to extract functions, subsequently reducing the image size at the pooling levels. The following blocks maintain a similar architecture, with blocks 1 and 2 employing two convolutional layers, while blocks 3—5 use three convolutional layers. In these later blocks, different kernel numbers on each layer deepen the network and enhance accuracy. Finally, three fully connected layers converge to output objects into two classifications.

The VGG16 structure encounters several challenges when applied to medical image datasets (Cheng et al., 2022). Originally designed to recognize 1000 categories from 1 million images, VGG16 may face limitations when applied to smaller datasets with fewer learning opportunities, resulting in inadequate training. This limitation becomes apparent when attempting to discern the presence or absence of pneumonia in chest X-rays, where the homogeneous nature of image characteristics makes it challenging for models to capture relevant features. Consequently, training with deeper network levels can lead to recognition errors (Jiang et al., 2021).

To address these issues, the proposed architecture, depicted in Fig. 8.6, aims to mitigate problems related to both under- and overtraining by reducing the network's depth compared with the original VGG16. The original convolutional architecture of VGG16 is maintained, but feature extraction is accomplished using two consecutive small convolutional kernels rather than a single large one. This approach retains the feature-highlighting effect of VGG16 while simultaneously reducing the number of parameters. This not only shortens training time but also preserves the depth of the network layer.

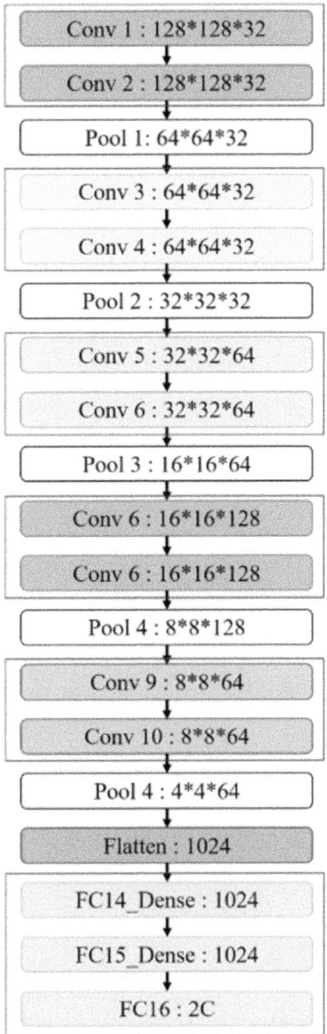

Figure 8.6 Proposed architecture of the modified VGG-16 network.

Training, validation and evaluation of models

To train and validate the models, we use a standard deep neural network training process. Below, we describe the training and validation process, as well as the metrics used to evaluate the performance of the models.

Each of the two models is trained on training data. During training, the models adapt to the input images and the corresponding class labels (pneumonia or normal lungs, presence or absence of a tumour). Model training is typically conducted over various numbers of epochs using the backpropagation method to find the optimal parameters for the model's learning.

After each epoch of model training, a validation dataset is used to evaluate the performance of the models and avoid overtraining. Validation metrics (Orozco-Arias et al., 2020), such as accuracy, sensitivity, specificity, and F1-score (Hicks et al., 2022), help determine how well a model classifies pneumonia and normal lung conditions. However, in the case of an unbalanced dataset, the F1-score is the best measure of model performance (Izonin et al., 2022).

In addition to standard metrics that evaluate classification accuracy, we also investigate the explanatory power of models. Explanability is important in medical AI applications as it allows us to understand which features in the images influence the model's decisions. We use methods such as LIME (Local Interpretable Model-agnostic Explanations) and SHAP (SHapley Additive exPlanations) to analyze the importance of features and explain the decisions made by the model (Rao et al., 2022).

This training and validation process is aimed at developing reliable and efficient models for classifying pneumonia and normal lung conditions from medical images. We continuously evaluate the performance and explanatory power of the models to ensure their reliability and applicability in medical practice.

Modeling and results

Model development and analysis tools

During the scientific research focused on the classification of a patient's condition based on images from computed tomography scans of the lungs and brain, with a focus on highlighting affected areas, a series of tools were employed for the development and analysis of deep learning models. One of the key components of these tools is the Python programming language, chosen as the primary development tool due to its popularity and rich ecosystem of libraries for machine learning. The use of Python simplifies the process of developing and researching models, enabling the rapid implementation of cutting-edge approaches in the field of deep learning.

In particular, the TensorFlow and Keras libraries were used to build and configure deep neural networks. TensorFlow provides a powerful framework for working with neural networks, and Keras allows developing models using a higher level of abstraction, which simplifies the process.

These libraries play a crucial role in creating and training models capable of classifying pneumonia based on the analysis of medical images of the lungs.

In parallel, for data processing and metric computation, the scikit-learn library was employed. It provides convenient tools for data preprocessing and model performance evaluation, facilitating the analysis of research results. By using these development tools, it becomes more efficient to explore and develop models for pneumonia classification based on medical images of the lungs, ensuring accuracy and reliability in the crucial medical context.

For training, an NVIDIA RTX3080 graphics processor with 10 gigabytes of video memory, CUDA technology (Mochurad, 2022), and the Linux operating subsystem to run the Jupyter Lab server environment for model training were utilized. Additionally, for comparison, training will also be conducted on an Intel Core i7 12700K processor. This allows evaluating the difference in accuracy without using CUDA technology.

Results of applying the proposed algorithm for Dataset 1

The classification quality investigation will be conducted by comparing the performance of each model individually and in the ensemble. For each step, a prepared dataset will be used.

Also, to understand the results, each model is wrapped in a LIME, a class that generates interpretable explanations for the model's solutions, even if the model itself is complex. This is done by training a locally interpreted model based on a subsample of the data.

Firstly, we consider the training result of the proposed deep convolutional network using the chest computed tomography dataset. We used 100 epochs for training and the EarlyStopping Keras method to stop the model training early.

The model training was stopped after 40 epochs, the total training time was 144 s, and an accuracy of 95.2% was achieved, with an average loss function value of 0.29357. The precision and recall metrics are 81% and 67%, respectively, and the f1-score is 73% (see Fig. 8.7).

Next, we will train the model with the proposed modified VGG16 network structure. The training will also be performed until an early stopping point with the maximum number of training epochs equal to 100.

The model reached the training stop at 24 epochs in 77 s with the accuracy and loss function values equal to 92.3% and 0.9567, respectively

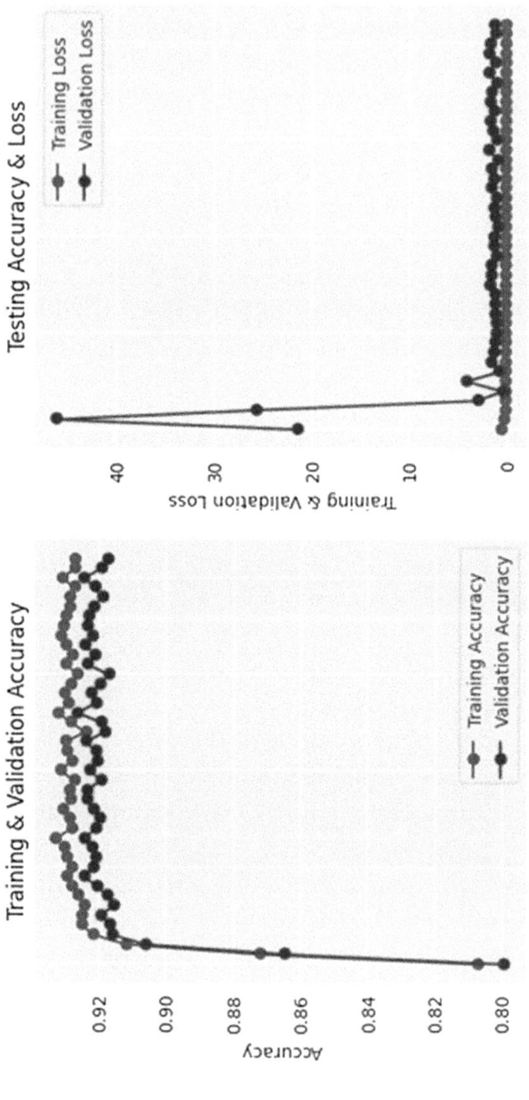

Figure 8.7 Accuracy and loss functions at each epoch of deep convolutional network training using a Dataset 2.

(see Fig. 8.8). The precision and recall metrics are 78% and 75%, respectively, and the f1-score is 77%.

By combining two models into an ensemble using the "majority voting" methodology, we will assess the classification accuracy and compare it with the performance of each network individually. Additionally, in the comparison, we will evaluate the models' performance using the Matthew's correlation coefficient (MCC) and Cohen's Kappa (CK) metrics. Since the f1-score is less truthful and informative than the coefficients mentioned above in binary classification assessments, these coefficients provide a more balanced evaluation.

The f1-score metric ignores true negative classifications, thus, it is not an ideal assessment for imbalanced classes, whereas the mentioned coefficients are symmetric and assess both directions of predictability — the classifier predicts the true class, and the true class predicts the classifier's prediction, offering separate multiclass indicators of accuracy and recall; the results are presented in Table 8.1.

Results of applying the proposed algorithm for Dataset 2

We will use a dataset of human brain computed tomography (CT) images to train the models. Since we normalized the image size during the previous data preparation, we will not alter the network parameters. We achieved an accuracy of 95.9% with the model, and it scored precision and recall metrics of 88% and 71% respectively, along with an f1-score of 78%. We stopped the model's training at the 50th epoch, which took 201 s (see Fig. 8.9).

We trained the network with the proposed modified VGG16 structure, resulting in a model with an accuracy of 92.3%, completing its training in 41 epochs and taking 150 s. This model achieved precision and recall metrics of 78% and 76%, respectively, with an f1-score of 77% (see Fig. 8.10).

Now, we will evaluate the ensemble's performance and compare it with the metrics obtained from each model's separate evaluation (see Table 8.2).

Analyzing the behavior of classification models when applying XAI methods

Let's consider the results of the obtained models and compare them with similar studies. To do this, we will examine the discrepancy matrices for each model and apply the methods of XAI to better understand the model's behavior during classification.

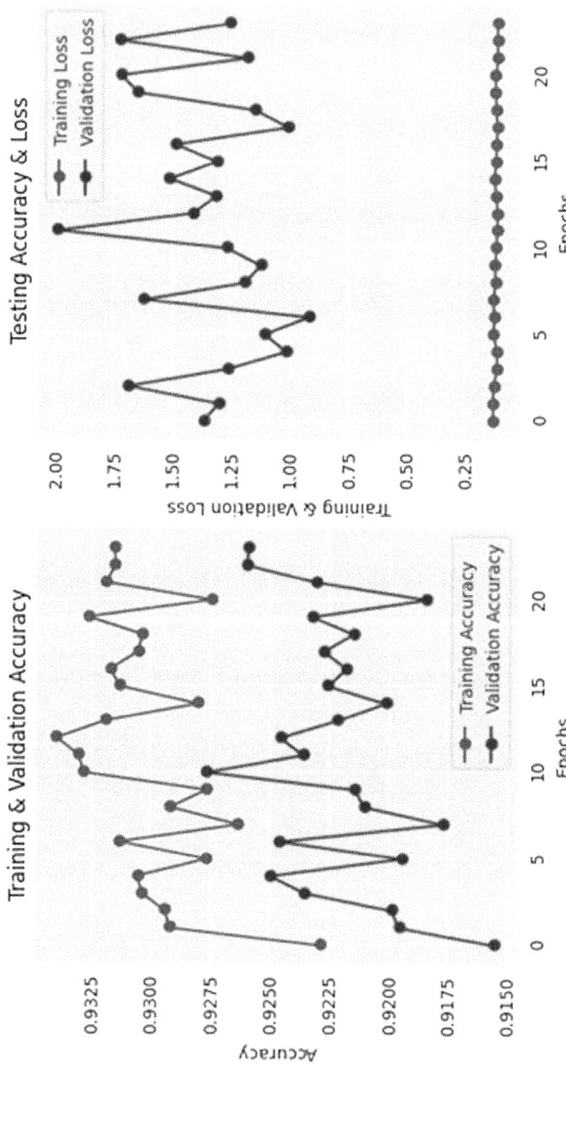

Figure 8.8 Accuracy indicators and loss functions at each epoch of model training with the proposed modified VGG16 network structure using a Dataset 1.

Table 8.1 Comparison of classification quality for the Dataset 1.

Model	Precision %	Recall %	F1-score %	MCC	CK
Deep CNN	81	67	73	0.78	0.74
Modified VGG16	78	75	77	0.82	0.76
Ensemble	83	76	79	0.84	0.80

Discussion results for Dataset 1

One of the challenges is the quantitative imbalance in the input images, where the "Pneumonia" class outnumbers the "Normal" class by more than threefold. This imbalance increases the likelihood of the model favoring the classification into the majority class.

In situations involving imbalanced datasets, accuracy may not be the most informative metric. For instance, in a dataset with 95 negative and 5 positive results, a model achieving 95% accuracy could simply label every example as negative. Hence, alternative metrics such as precision and recall become crucial for a more meaningful evaluation of models dealing with such imbalances.

To address this, let's first examine the confusion matrices for the models trained using the lung computed tomography (CT) image dataset (refer to Figs. 8.11–8.13).

Since classification accuracy is not the best metric for an imbalanced dataset, the confusion matrices in Fig. 8.11 allow us to see the distribution of correct and incorrect image classifications relative to their class.

The biggest issue with the proposed deep convolutional network is the high number of false negative predictions. This is not apparent when determining the model's accuracy, as there are fewer images of healthy patients, but it can be critical when using the model in real-world conditions.

The model using the proposed modified VGG16 structure shows a more balanced distribution of false and true predictions, as presented in Fig. 8.12. Overall, this reflects its lower accuracy, but in practice, the ratio of true negatives to false negatives is better than in the first model, although it is less effective in the ratio of true positives to false positives.

The best ratios of true to false image classifications were achieved by the ensemble (see Fig. 8.13), which leveraged the strengths of each model. This

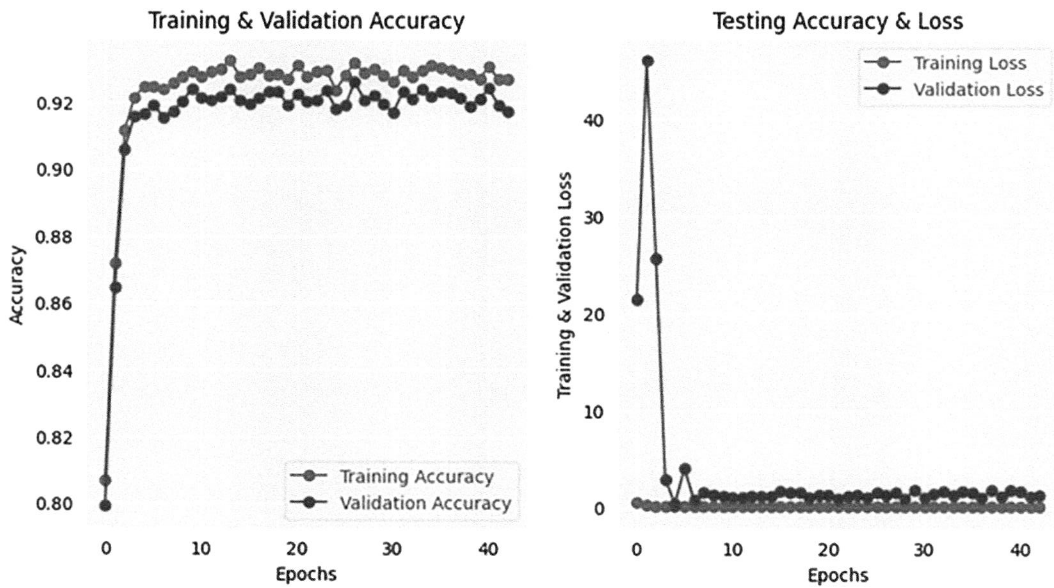

Figure 8.9 Accuracy and loss functions at each epoch of deep convolutional network training using a Dataset 2.

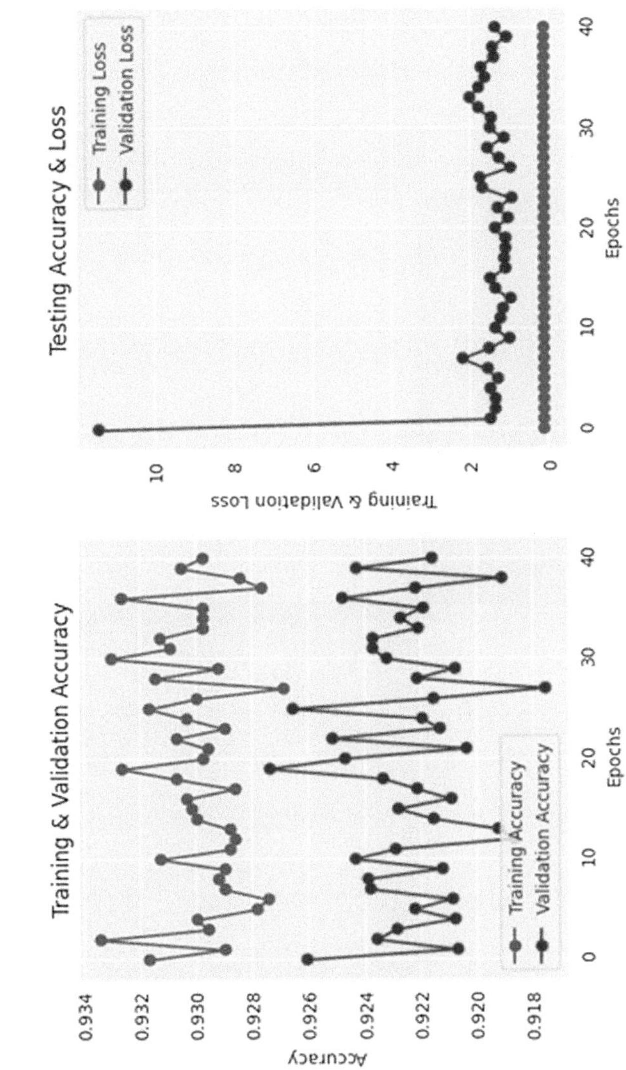

Figure 8.10 Accuracy indicators and loss functions at each epoch of model training with the proposed modified VGG16 network structure using a Dataset 2.

Table 8.2 Comparison of classification quality for the Dataset 2.

Model	Precision %	Recall %	F1-score %	MCC	CK
Deep CNN	88	71	78	0.81	0.77
Modified VGG16	78	76	79	0.84	0.76
Ensemble	82	79	80	0.86	0.82

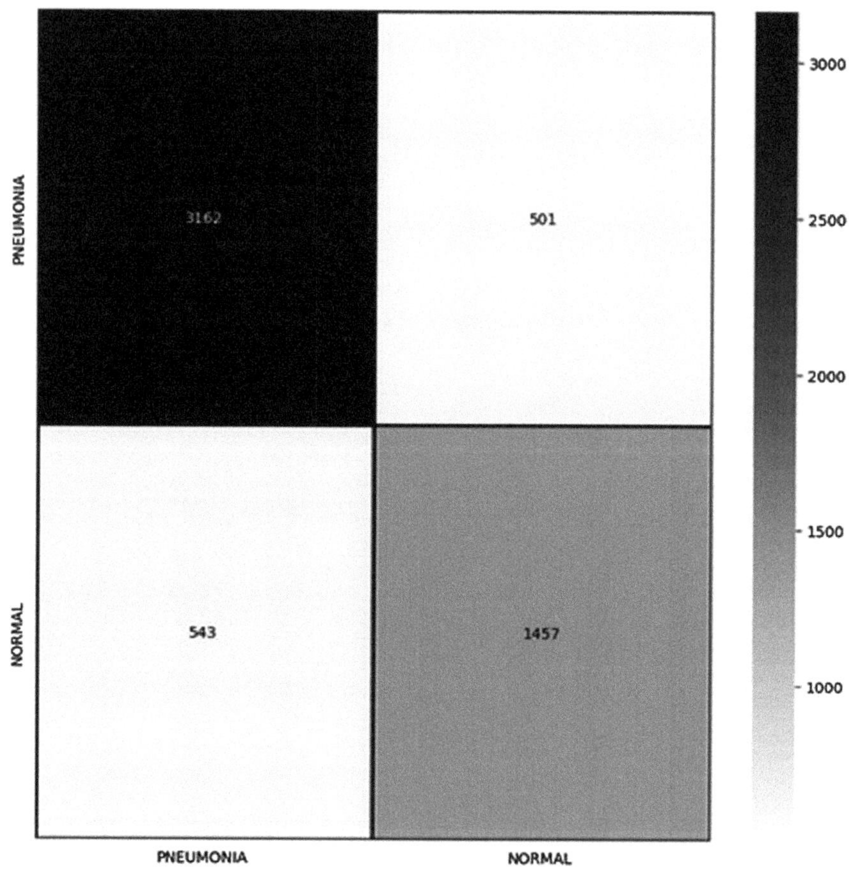

Figure 8.11 Confusion matrices for the trained deep convolutional network model for Dataset 1.

allowed for the most optimal ratios and helped avoid misclassifications due to potential overfitting or undertraining of any single model.

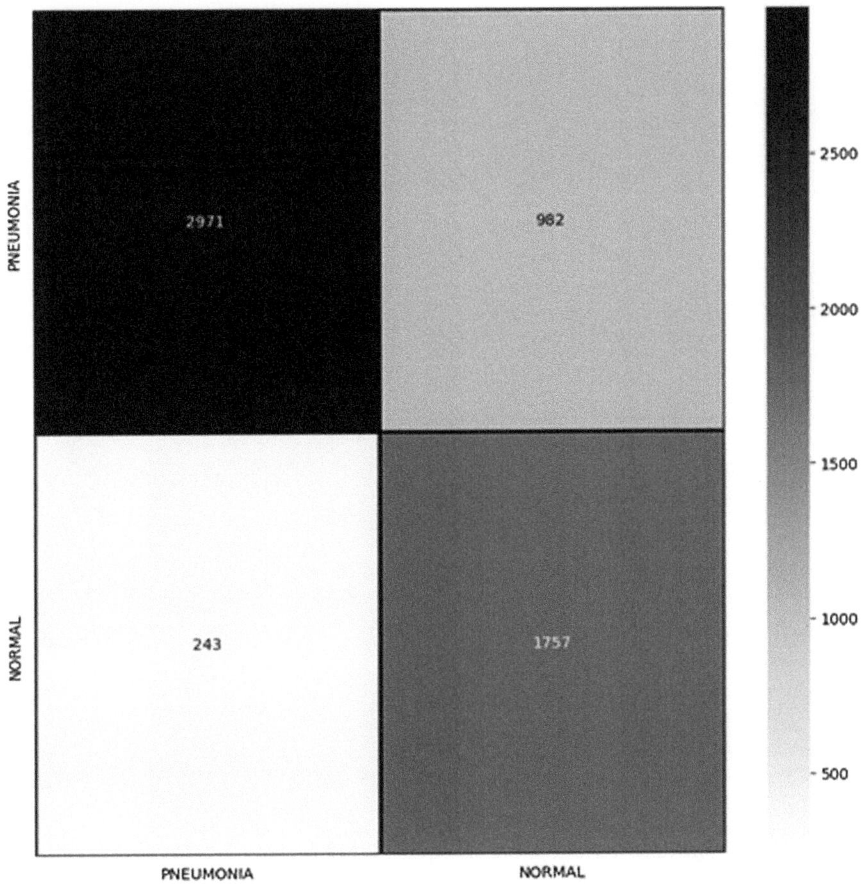

Figure 8.12 Confusion matrices for the trained VGG16 model with modified structure for Dataset 1.

To understand the results obtained from our models, we will apply feature visualization, which was decisive in classifying each image. We'll start with the explanation of results from the deep CNN model (see Fig. 8.14). We identified feature maps using the Explainer class from libraries such as LIME and SHAP. The highlighted features were approximated with increased smoothing to obtain generalized zones of images that the models used for making classification decisions.

The concept of explanation involves providing a local linear approximation of the model's behavior. Even if a model is highly complex on a global scale, it is more manageable to approximate its behavior in the proximity of a specific instance. When treating the model as a black box,

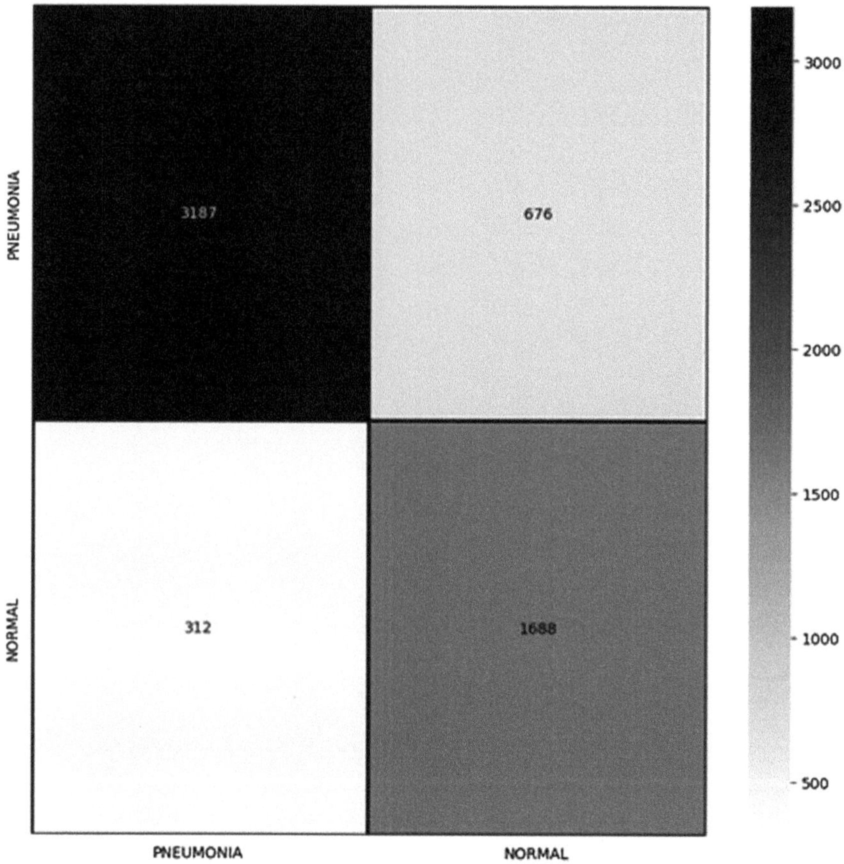

Figure 8.13 Confusion matrices for the ensemble of trained networks for Dataset 1.

the approach is to perturb the instance targeted for explanation, and then learn a sparse linear model around it to serve as an explanation.

The given Fig. 8.14 illustrates the intuition of this procedure. The decision-making function of the model is represented in red. The area marked in red had signs of pneumonia in the patient. In this example, the area on the left had more weight, so the model classified the presence of pneumonia. Green occupied a smaller area. However, the model took into account a rather large area of the feature, which was responsible for the class of a healthy patient. This is a case of overtraining, causing the model to make many misclassifications of the healthy patient.

Our second model in feature isolation had a higher degree of tolerance in feature classification, which allowed us to reduce the number of

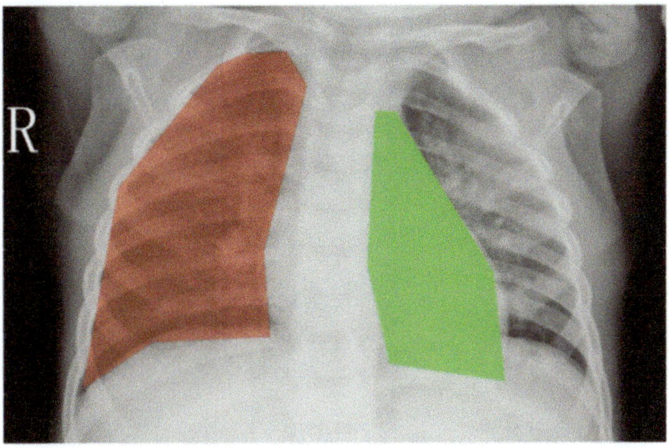

Figure 8.14 Image of key features in the classification of pneumonia based on the deep CNN model.

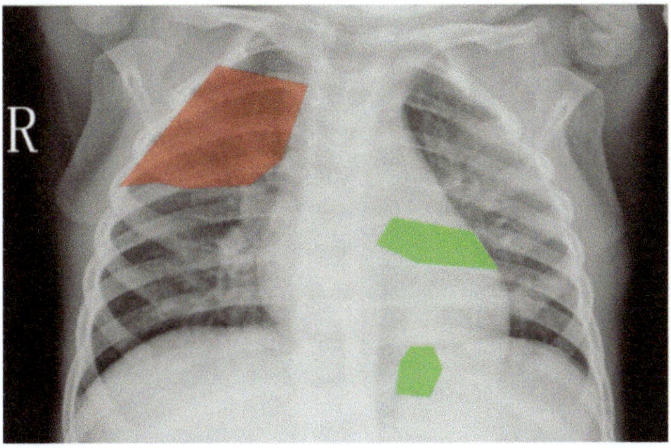

Figure 8.15 Image of key features in the classification of pneumonia based on the VGG-16 convolutional network with modification.

misclassified healthy patients (see Fig. 8.15). However, the plane, which was responsible for the key feature of the sick patient, had a significantly smaller size, which is explained by a higher percentage of falsely classified sick patients than in the first model.

The highest quality of predictions was obtained using the proposed ensemble (see the results in Fig. 8.16). The weight of the feature of the sick patient in this case was used by the first model, while the feature that was responsible for classifying the patient as healthy had a lower weight.

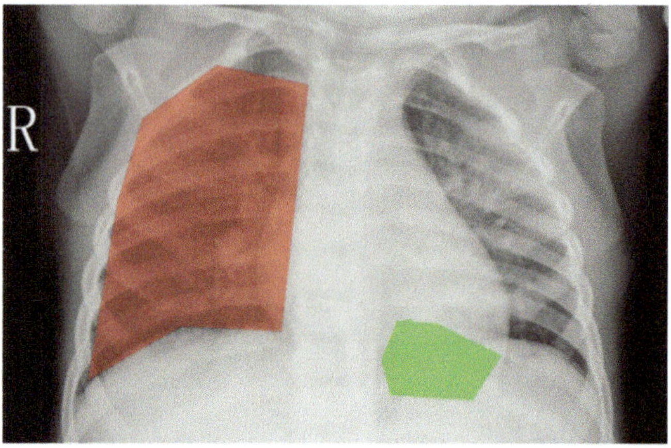

Figure 8.16 Image of key features in the classification of pneumonia based on the proposed ensemble.

Considering the peculiarity of the problem, the priority in this task is the classification with a higher probability of the pneumonia class. With the help of XAI, the user is provided with an explanation of the choice of one or another class, by highlighting features on the image.

To compare the difference between the approaches, it is worth mentioning the work (Ahmed et al., 2019), in which, with the help of approaches of logical methods of image analysis, in which, for the sake of balance, the accuracy and response values were almost the same, an accuracy of 87% was obtained for the detection of the key feature and 83% for the final classification.

Discussion results for Dataset 2

Next, let's examine the results of training the models on the second dataset — CT scans of the human brain. This dataset did not have the issue of imbalance, allowing us to explore how the networks make decisions based on the patient's state.

First, let's consider the confusion matrix for the deep convolutional network (see Fig. 8.17).

In Fig. 8.18, the red polygons represent regions of the image that highlight a feature classified as a tumor. Features highlighted in green represent areas where the probability of belonging to the tumor class was insufficient for classification as a tumor.

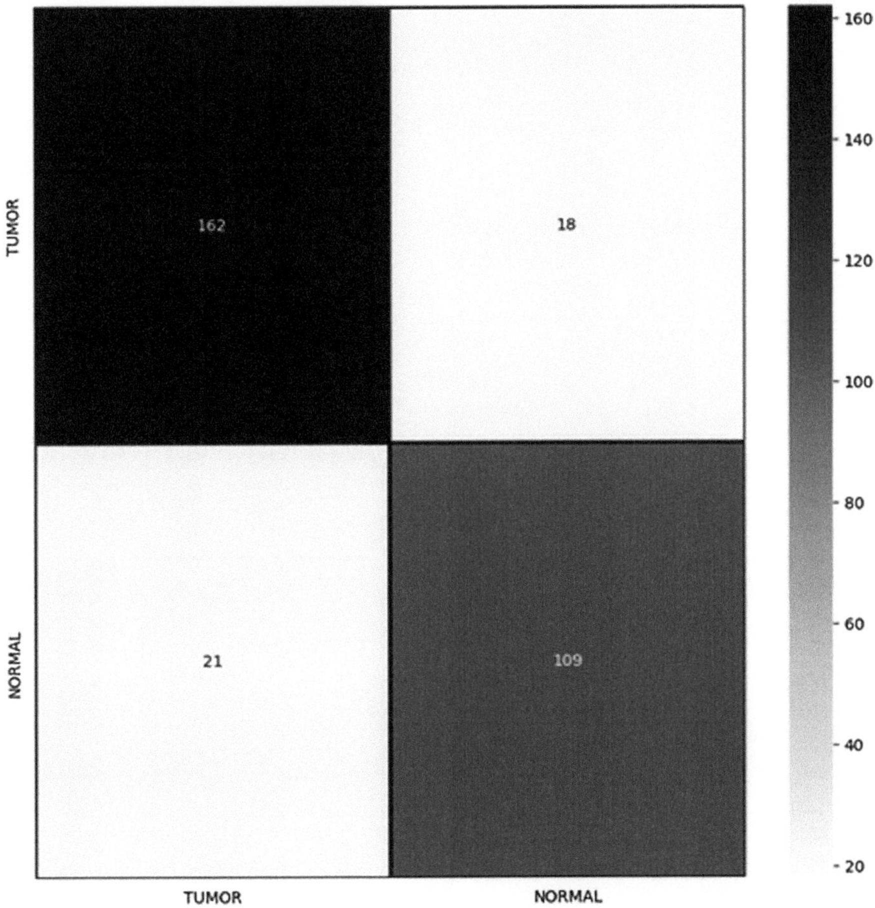

Figure 8.17 Confusion matrix for the deep convolutional network for Dataset 2.

When highlighting features such as tumors, it is important to consider that the size and appearance of malignant tumors can sometimes vary. Although in this case, these false tumors were rejected by the detection threshold, there is still a certain number of cases where genuine tumors were not classified.

The modified VGG16 showed low rates of misclassified patients with existing brain tumors. However, there is also a significant number of healthy patients who were misclassified (see Fig. 8.19). This is further illustrated using XAI in Fig. 8.20.

It is worth noting the absence of probable tumors and a smaller highlighted area of the tumor. This can be attributed to the higher threshold

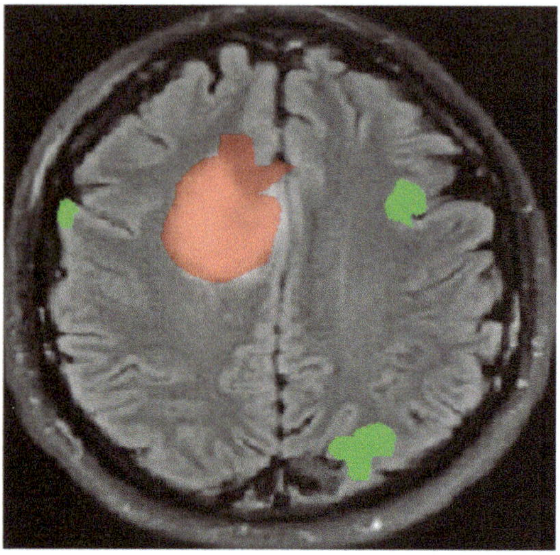

Figure 8.18 Images of key features during brain tumor classification for Dataset 2.

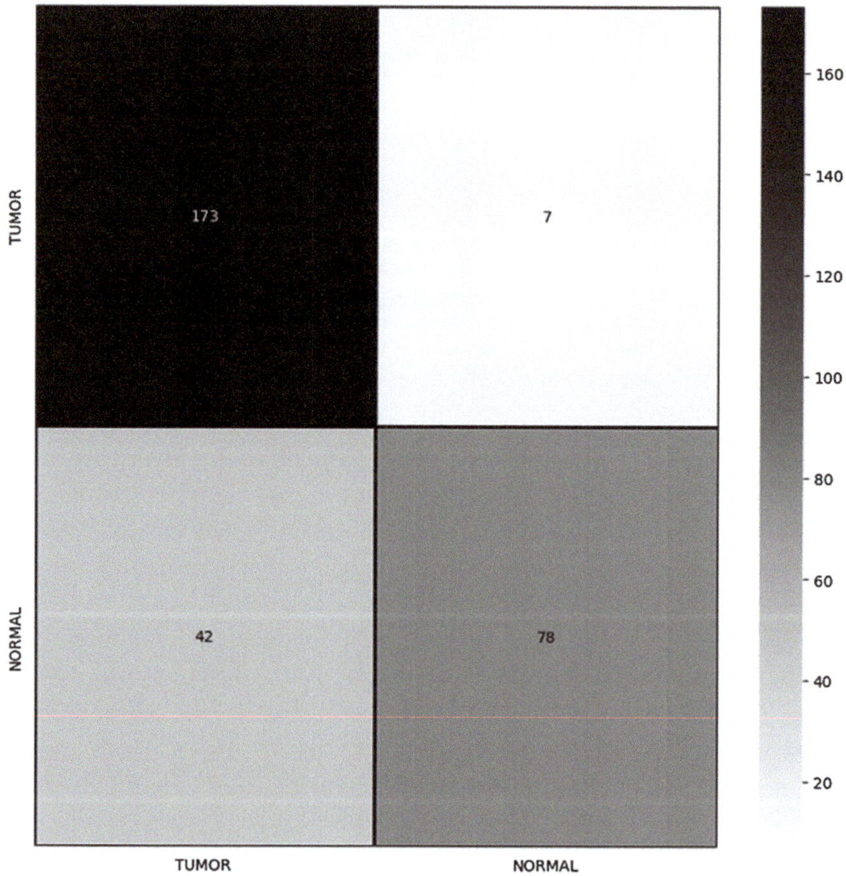

Figure 8.19 Confusion matrix for the modified VGG16 for Dataset 2.

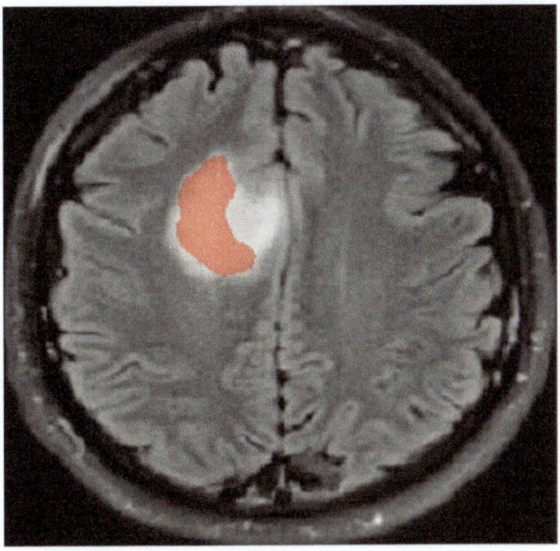

Figure 8.20 Images of key features during brain tumor classification.

chosen by the model during training on the dataset. However, this also affected how many input images without tumors were classified as having tumors.

The proposed ensemble combining two networks improved the classification of healthy patients while maintaining the tumor detection ratio (see Fig. 8.21). This is explained by the fact that, during the classification of images with tumors, the weight of decisions from the second network was higher. This allowed an increase in the number of true positive predictions, while the first network helped reduce the number of false negatives.

In Fig. 8.22, we observe that although the area of the tumor has increased, the network did not highlight any false regions as probable tumors.

Having analyzed the results, we found that due to the imbalanced dataset, there is an increased likelihood of classifying images as having features of pneumonia and brain tumors. The use of an ensemble of networks improved the f1-score metric, providing evidence of the effectiveness of the network ensemble.

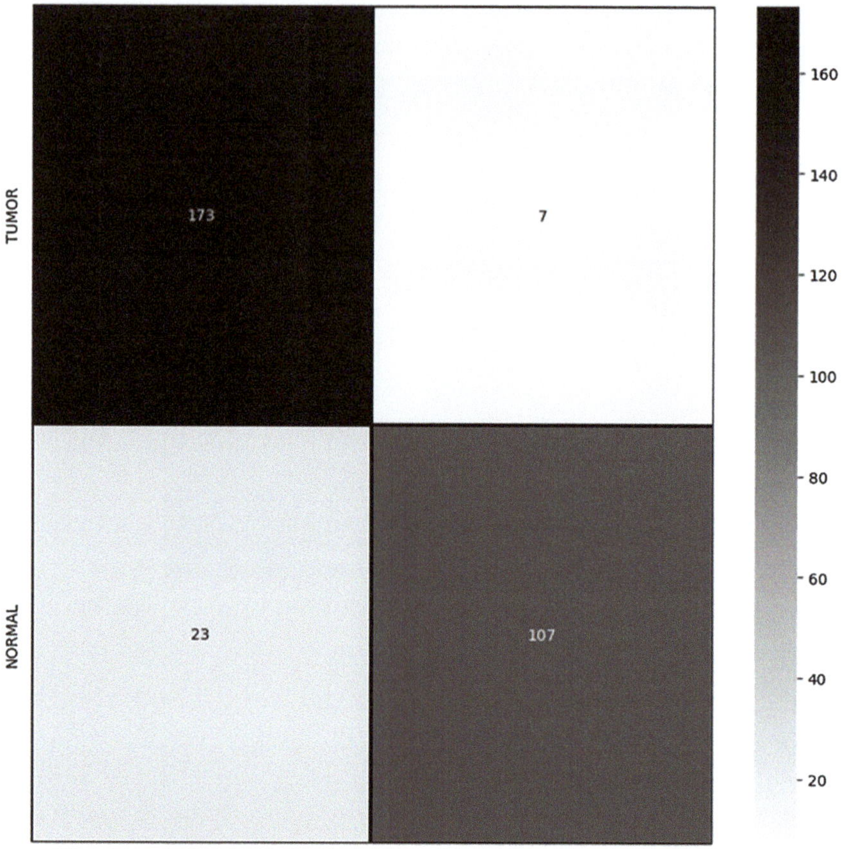

Figure 8.21 Confusion matrix for the proposed ensemble for Dataset 2.

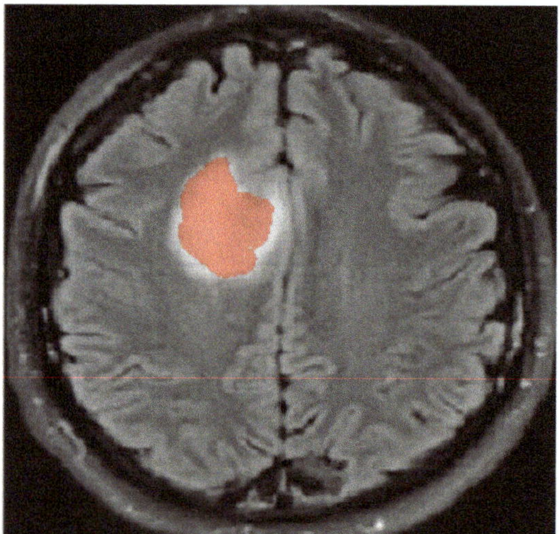

Figure 8.22 Images of key features during brain tumor classification.

Conclusions

The work established a medical image analysis system employing deep learning, specifically leveraging the proposed ensemble of neural networks. Additionally, it identified key areas of focus within the realm of medical analysis.

The issue of data processing and preparation for image analysis for the purpose of classification using deep learning is considered. Necessary steps for preliminary data preparation such as normalization, resizing, denoising, and data augmentation are described. Deep neural network structure was provided, for classification based on medical images. The process of training and validating the models is described, including the use of metrics to evaluate productivity and techniques for XAI that are important for understanding the decisions made by the models and that help interpret the results.

The resulting model trained on a deep convolutional network was used in conjunction with the LIME and SHAP model explanation algorithms, which made it possible to explain the model solution by highlighting the features. After analyzing the results, it was found that due to an unbalanced data set, there is an increased probability of classifying the image as showing signs of pneumonia. Also, for the possibility of making a decision in real time, it is proposed to use such parallel computing technology as CUDA.

So, as a result, the modified VGG16 model showed an improvement in f1-score by 4% for the first dataset. And the proposed ensemble made it possible to improve this indicator by 6%. For the second dataset, we have an improvement of 1% and 2%, respectively.

As further research, it would be important to analyze the possibilities of alternative ways of functional parallelization (Mochurad, 2021; Mochurad et al., 2021) of the proposed computer tomography image classification algorithm using an ensemble of neural networks, as well as to conduct additional studies on datasets of larger dimensions.

Funding

This research was funded by National Research Foundation of Ukraine project number 2023.03/0029.

Acknowledgments

The National Research Foundation of Ukraine funds this study from the state budget of Ukraine within the project No2023.03/0029.

References

Ahmed, A.M., Ibrahim, S.K., Yacout, S., 2019. Hyperspectral image classification based on logical analysis of data. In: IEEE Aerospace Conference Proceedings, vol 2019. IEEE Computer Society, Egypt. https://doi.org/10.1109/AERO.2019.8742023, 9781538668542.

Altameem, A., Kovtun, V., Al-Ma'aitah, M., Altameem, T., Fouad, H., Youssef, A.E., 2022. Patient's data privacy protection in medical healthcare transmission services using back propagation learning. Comput. Electr. Eng. 102, 108087. https://doi.org/10.1016/j.compeleceng.2022.108087, 00457906.

Bo, Z.H., Guo, Y., Lyu, J., Liang, H., He, J., Deng, S., Xu, F., Lou, X., Dai, Q., 2023. Relay learning: a physically secure framework for clinical multi-site deep learning. Npj Dig. Med. 6 (1). https://doi.org/10.1038/s41746-023-00934-4, 23986352.

Bortsova, G., González-Gonzalo, C., Wetstein, S.C., Dubost, F., Katramados, I., Hogeweg, L., Liefers, B., van Ginneken, B., Pluim, J.P.W., Veta, M., Sánchez, C.I., de Bruijne, M., 2021. Adversarial attack vulnerability of medical image analysis systems: unexplored factors. Med. Image Anal. 73. https://doi.org/10.1016/j.media.2021.102141, 13618415.

Brain Tumor, 2020. https://www.kaggle.com/datasets/jakeshbohaju/brain-tumor. (Accessed 24 November 2023).

Chaitanya, K., Karani, N., Baumgartner, C.F., Erdil, E., Becker, A., Donati, O., Konukoglu, E., 2021. Semi-supervised task-driven data augmentation for medical image segmentation. Med. Image Anal. 68. https://doi.org/10.1016/j.media.2020.101934, 13618415.

Cheng, J., Song, Q., Peng, H., Huang, J., Wu, H., Jia, B., 2022. Optimization of VGG16 algorithm pattern recognition for signals of michelson−sagnac interference vibration sensing system. Photonics 9 (8). https://doi.org/10.3390/photonics9080535, 2304−6732.

Chest X-Ray Images (Pneumonia), 2018. https://www.kaggle.com/datasets/paultimothymooney/chest-xray-pneumonia. (Accessed 24 November 2023).

Chumachenko, D., Dudkina, T., Chumachenko, T., Morita, P.P., 2023. Epidemiological implications of war: machine learning estimations of the Russian invasion's effect on Italy's COVID-19 dynamics. Computation 11 (11), 221. https://doi.org/10.3390/computation11110221, 2079−3197.

Goyal, L., Dhull, A., Singh, A., Kukreja, S., Singh, K.K., 2022. VGG-COVIDNet: a Novel model for COVID detection from X-Ray and CT scan images. Procedia Computer Science 218, 1926−1935. https://doi.org/10.1016/j.procs.2023.01.169, 18770509.

He, S., Pereira, D., David Perez, J., Gollub, R.L., Murphy, S.N., Prabhu, S., Pienaar, R., Robertson, R.L., Ellen Grant, P., Ou, Y., 2021. Multi-channel attention-fusion neural network for brain age estimation: accuracy, generality, and interpretation with 16,705 healthy MRIs across lifespan. Med. Image Anal. 72. https://doi.org/10.1016/j.media.2021.102091, 13618415.

Hicks, S.A., Strümke, I., Thambawita, V., Hammou, M., Riegler, M.A., Halvorsen, P., Parasa, S., 2022. On evaluation metrics for medical applications of artificial intelligence. Sci. Rep. 12 (1). https://doi.org/10.1038/s41598-022-09954-8, 20452322.

Hussain, S., Lafarga-Osuna, Y., Ali, M., Naseem, U., Ahmed, M., Tamez-Peña, J.G., 2023. Deep learning, radiomics and radiogenomics applications in the digital breast tomosynthesis: a systematic review. BMC Bioinf. 24 (1). https://doi.org/10.1186/s12859-023-05515-6, 14712105.

Hyun, C.M., Baek, S.H., Lee, M., Lee, S.M., Seo, J.K., 2021. Deep learning-based solvability of underdetermined inverse problems in medical imaging. Med. Image Anal. 69. https://doi.org/10.1016/j.media.2021.101967, 13618423.

Izonin, I., Tkachenko, R., Greguš, M., 2022. I-PNN: an improved probabilistic neural network for binary classification of imbalanced medical data. In: Lecture Notes in

Computer Science (Including Subseries Lecture Notes in Artificial Intelligence and Lecture Notes in Bioinformatics), vol 13427. Springer Science and Business Media Deutschland GmbH, Ukraine, pp. 147–157. https://doi.org/10.1007/978-3-031-12426-6_12, 16113349.

Jangra, M., Dhull, S.K., Singh, K.K., Singh, A., 2023. Denoising techniques for ECG arrhythmia classification systems: an experimental approach. In: Lecture Notes on Data Engineering and Communications Technologies, vol 175. Springer Science and Business Media Deutschland GmbH, India, pp. 1–13. https://doi.org/10.1007/978-981-99-3432-4_1, 23674520.

Jiang, Z.P., Liu, Y.Y., Shao, Z.E., Huang, K.W., 2021. An improved VGG16 model for pneumonia image classification. Appl. Sci. 11 (23). https://doi.org/10.3390/app112311185, 20763417.

Jin, C., Udupa, J.K., Zhao, L., Tong, Y., Odhner, D., Pednekar, G., Nag, S., Lewis, S., Poole, N., Mannikeri, S., Govindasamy, S., Singh, A., Camaratta, J., Owens, S., Torigian, D.A., 2022. Object recognition in medical images via anatomy-guided deep learning. Med. Image Anal. 81, 102527. https://doi.org/10.1016/j.media.2022.102527, 13618415.

Liu, J., Dong, B., Wang, S., Cui, H., Fan, D.-P., Ma, J., Chen, G., 2021. COVID-19 lung infection segmentation with a novel two-stage cross-domain transfer learning framework. Med. Image Anal. 74. https://doi.org/10.1016/j.media.2021.102205, 13618415.

Mahapatra, D., Poellinger, A., Reyes, M., 2022. Interpretability-guided inductive bias for deep learning based medical image. Med. Image Anal. 81. https://doi.org/10.1016/j.media.2022.102551, 13618415.

Mochurad, L.I., 2022. Canny edge detection analysis based on parallel algorithm, constructed complexity scale and CUDA. Comput. Inf. 41 (4), 957–980. https://doi.org/10.31577/cai_2022_4_957, 2585–8807.

Mochurad, L., 2021. Optimization of regression analysis by conducting parallel calculations. CEUR Workshop Proceedings 2870, 982–996, 16130073.

Mochurad, L., Dereviannyi, A., Antoniv, U., 2021. Classification of X-ray images of the chest using convolutional neural networks. In: CEUR Workshop Proceedings, pp. 269–282, 3038, 16130073.

Mochurad, L., Panto, R., 2023. A parallel algorithm for the detection of eye disease. In: Lecture Notes on Data Engineering and Communications Technologies, vol 158. Springer Science and Business Media Deutschland GmbH, Ukraine, pp. 111–125. https://doi.org/10.1007/978-3-031-24475-9_10, 23674520.

Mondal, S.S., Mandal, N., Singh, K.K., Singh, A., Izonin, I., 2023. EDLDR: an ensemble deep learning technique for detection and classification of diabetic retinopathy. Diagnostics 13 (1). https://doi.org/10.3390/diagnostics13010124, 20754418.

Orozco-Arias, S., Piña, J.S., Tabares-Soto, R., Castillo-Ossa, L.F., Guyot, R., Isaza, G., 2020. Measuring performance metrics of machine learning algorithms for detecting and classifying transposable elements. Processes 8 (6). https://doi.org/10.3390/pr8060638, 2227–9717.

Paithane, P., Kakarwal, S., 2023. LMNS-Net: lightweight Multiscale Novel Semantic-Net deep learning approach used for automatic pancreas image segmentation in CT scan images. Expert Syst. Appl. 234. https://doi.org/10.1016/j.eswa.2023.121064, 09574174.

Pereira, P.M.., Thomaz, L.A., Tavora, L.M.., Assuncao, P.A.., Fonseca-Pinto, R.M., Paiva, R.P., Faria, S.M.M.d., 2022. Melanoma classification using light-Fields with morlet scattering transform and CNN: surface depth as a valuable tool to increase detection rate. Med. Image Anal. 75. https://doi.org/10.1016/j.media.2021.102254, 13618423.

Rakhra, A., Gupta, R., Singh, A., 2021. Medical Imaging and Diagnosis Using Machine Learning and Deep Learning. Informa UK Limited, pp. 390−394. https://doi.org/10.1201/9781003193838-71, 9781003193838.

Rao, S., Mehta, S., Kulkarni, S., Dalvi, H., Katre, N., Narvekar, M., 2022. A study of LIME and SHAP model explainers for autonomous disease predictions. In: IBSSC 2022 - IEEE Bombay Section Signature Conference. Institute of Electrical and Electronics Engineers Inc., India https://doi.org/10.1109/IBSSC56953.2022.10037324, 9781665492911.

Sharma, S., Guleria, K., 2023. A comprehensive review on federated learning based models for healthcare applications. Artif. Intell. Med. 146. https://doi.org/10.1016/j.artmed.2023.102691, 09333657.

Singh, A., Kant Singh, K., 2024. Blockchain and deep learning: research challenges, open problems, and future. In: Blockchain and Deep Learning for Smart Healthcare. CRC Press, India, pp. 425−431, 9781119792406. https://www.wiley.com/en-se/Blockchain+and+Deep+Learning+for+Smart+Healthcare-p-9781119792383.

Sun, H., Yang, S., Chen, L., Liao, P., Liu, X., Liu, Y., Wang, N., 2023. Brain tumor image segmentation based on improved FPN. BMC Med. Imag. 23 (1). https://doi.org/10.1186/s12880-023-01131-1, 1471−2342.

Wang, R., Chen, S., Ji, C., Fan, J., Li, Y., 2022. Boundary-aware context neural network for medical image segmentation. Med. Image Anal. 78. https://doi.org/10.1016/j.media.2022.102395, 13618415.

Wang, X., Yuan, Y., Guo, D., Huang, X., Cui, Y., Xia, M., Wang, Z., Bai, C., Chen, S., 2022. SSA-Net: spatial self-attention network for COVID-19 pneumonia infection segmentation with semi-supervised few-shot learning. Med. Image Anal. 79. https://doi.org/10.1016/j.media.2022.102459, 13618415.

Zhang, L., Wang, X., Yang, D., Sanford, T., Harmon, S., Turkbey, B., Wood, B.J., Roth, H., Myronenko, A., Xu, D., Xu, Z., 2020. Generalizing deep learning for medical image segmentation to unseen domains via deep stacked transformation. IEEE Trans. Med. Imag. 39 (7), 2531−2540. https://doi.org/10.1109/TMI.2020.2973595, 1558254X.

Zhang, R., Chung, A.C.S., 2021. MedQ: lossless ultra-low-bit neural network quantization for medical image segmentation. Med. Image Anal. 73. https://doi.org/10.1016/j.media.2021.102200, 13618423.

Zhao, Y., Guo, Q., Zhang, Y., Zheng, J., Yang, Y., Du, X., Feng, H., Zhang, S., 2023. Application of deep learning for prediction of Alzheimer's disease in PET/MR imaging. Bioengineering 10 (10). https://doi.org/10.3390/bioengineering10101120, 2306−5354.

CHAPTER 9

Hybrid ensemble learning model to improve the performance and interpretability of medical diagnosis: Small data tasks

Ivan Izonin[1], Roman Tkachenko[2], Veronika Kachmar[1] and Myroslav Havryliuk[1]
[1]Lviv Polytechnic National University, Department of Artificial Intelligence, Lviv, Ukraine; [2]Lviv Polytechnic National University, Department of Publishing Information Technologies, Lviv, Ukraine

Introduction

The modern development of medicine, in particular medical diagnostics, has become largely dependent on achievements in the field of artificial intelligence (Berezsky et al., 2023). This was preceded by several reasons. In particular, the emergence of the Internet of Things and the rapid development of hardware made it possible to digitize the field of medicine, which made it possible to collect and store large amounts of historical data on the course of diseases, the effectiveness of their treatment methods, and information about patient support during rehabilitation (Goyal et al., 2022). Such various types of data could and should have been processed in some way (Hovorushchenko et al., 2021). Large volumes of such data, different types, and complex relationships between variables impose many restrictions on analysis by a person (Shakhovska et al., 2022). In this regard, AI has become a rather promising tool for performing intellectual analysis of such data.

The development of powerful deep learning tools (around 2010) ensured high processing efficiency for huge amounts of medical data (Saravanan et al., 2021). This made it possible to implement artificial neural networks (ANN) of deep learning in various components of medical systems for their intellectualization (Berezsky et al., 2019). However, the complexity of the learning algorithms of such AI tools, the huge number of

Responsible and Explainable Artificial Intelligence in Healthcare
ISBN 978-0-443-24788-0
https://doi.org/10.1016/B978-0-443-24788-0.00009-1

hyperparameters that need to be selected for the effective operation of ANNs, etc., provoked several problems, in particular, concerning understanding how they reach some conclusion. Their decisions were rather difficult to interpret (Krak et al., 2023). This caused medical workers to distrust the results of such systems, which in turn reduced the possibility of practical use.

The need to understand how these models make decisions led to the development of the field of explanatory AI (Manziuk et al., 2021). From around the 2010s onwards, scholars have actively begun to study and develop interpretability techniques, including visualization tools (Krak et al., 2022), explanation algorithms, and other approaches that allow you to understand what factors influence the decision-making process of models, and how the model in general reaches a particular conclusion. This direction became very relevant in medicine, where it was about human life and health.

Interpreted artificial intelligence (IAI) began to play a significant role in modern tasks of medical diagnosis, treatment, and rehabilitation (Nguyen Gia et al., 2019; Shafay et al., 2023). In particular, IAI began to be successfully used to analyze medical images (photographs, X-rays, MRI) and laboratory data for accurate diagnosis of diseases. Deep learning models began to be used to identify anomalies and pathologies (Radiuk et al., 2022), which helped doctors make quick and accurate decisions. IAI began to provide explanations for its findings, allowing doctors to better understand the models' recommendations and take them into account when making decisions about personalized treatment. In addition, AI began to be used to automate routine tasks, such as image analysis, verification of medical information, interpretation of test results, etc. This allowed doctors to focus on more complex tasks and improved the efficiency of the medical staff. Also, various IAI models have been used to predict disease risks based on patient history and other factors. However, only interpreted models were trusted and, accordingly, were successfully implemented in real medical systems. This helped to identify high-risk patients on time and assign preventive treatment. The use of IoT and interpreted IAI models have made it possible to monitor the patient's response to treatment and recommend corrections in real-time, which improves treatment outcomes.

In general, IAI in medical diagnostics can help combine the high accuracy of machine learning algorithms with the understanding and trust of medical professionals in these methods, making it a very relevant area of research and development.

However, the modern development of methods for interpreting highly effective models in the analysis of tabular datasets requires further refinement. On the one hand, simple linear models that are easy to interpret are not able to provide high classification accuracy. On the other hand, more complex models provide sufficient accuracy, but their results are difficult to explain. A contradiction arises between the need to ensure the highest possible accuracy of the model and the interpretability of the obtained results.

This is why this chapter aims to develop a new, ensemble-based machine learning model that will improve the efficiency of analysis of small datasets (accuracy, generalization, speed) the results of which can be easily interpreted.

The main contribution of this chapter is as follows:

- We designed a new ensemble-based method for small data classification based on the cascading of two machine learning methods, which provided both an increase in efficiency and the possibility of interpreting the result;
- We developed three different algorithmic implementations of the method to increase the accuracy and/or speed of processing data of different dimensions and with different number of classes to increase the efficiency of solving various medical diagnosis tasks;
- We compared the developed algorithmic implementations of the new cascade method with other methods based on the accuracy, speed, and interpretability level of their work, and showed the high efficiency of their application on three different datasets.

Background

Naïve Bayes

Naïve Bayes classifier (Naïve Bayes classifier) is one of the simplest statistical machine learning algorithms that uses the Bayes theorem to solve the problem of object classification. The title of the classifier includes the word "naive" because it is based on the assumption that all attributes of a given instance are independent.

The basis of the method is the procedure for assessing the probability of a specific instance belonging to each of the defined classes of the problem. Such probabilities are formed based on various independent attributes of each specific instance using Bayes' theorem. Mathematical explanations of this method are presented in (Vu, 2022).

Among the advantages of this method of data classification for solving the problem set in the paper, the following should be highlighted.

- Naïve Bayes is a relatively simple algorithm that demonstrates satisfactory classification accuracy in the case of a small amount of training data (small data samples) (Bisikalo et al., 2023);
- The naivety of the algorithm allows you to effectively work with a large number of features (high-dimensional datasets) (Pavliuk and Kolesnyk, 2023);
- It works well with sparse data, which is a typical problem in the tasks of processing samples of medical data;
- Naïve Bayes is especially effective for working with categorical attributes, which quite often arise when it is necessary to take into account different attributes of the object of attention during data collection

However, the main advantage of the Naïve Bayes classifier in the context of solving the stated task is the relative simplicity of interpreting the results of its work. It follows from the possibility of using probabilities as the results of its work, which the method returns for each specific instance. The maximum probability from the obtained ones determines whether the instance belongs to a specific class of the task.

Despite this, the main drawback of this machine learning method is its assumption about the independence of the attributes of the given object of attention. In the case of medical data analysis, high-dimensional datasets contain complex nonlinear relationships between the attributes of each instance, which can reduce the accuracy of their analysis using this method.

SGTM neural-like structures

The neural-like structure of the Successive Geometric Transformations Model (SGTM) is an analog of a neural network developed by Tkachenko (2022); Tkachenko and Izonin (2019). This innovative model uses a greedy noniterative learning algorithm that helps improve the performance of feed-forward neural networks for various data mining tasks.

Fig. 9.1 represents a basic, unsupervised version of the SGTM implementation. In this mode, the same signals are applied to the input and the output (Tkachenko, 2022; Tkachenko and Izonin, 2019).

The training mode of this SGTM version based on a sample of data, where each element of the matrix is labeled $X_{i,j}$, where $i = \overline{1, N}$, is the row number and $j = \overline{1, n}$, is the column number, requires the following steps (Tkachenko, 2022; Tkachenko and Izonin, 2019):

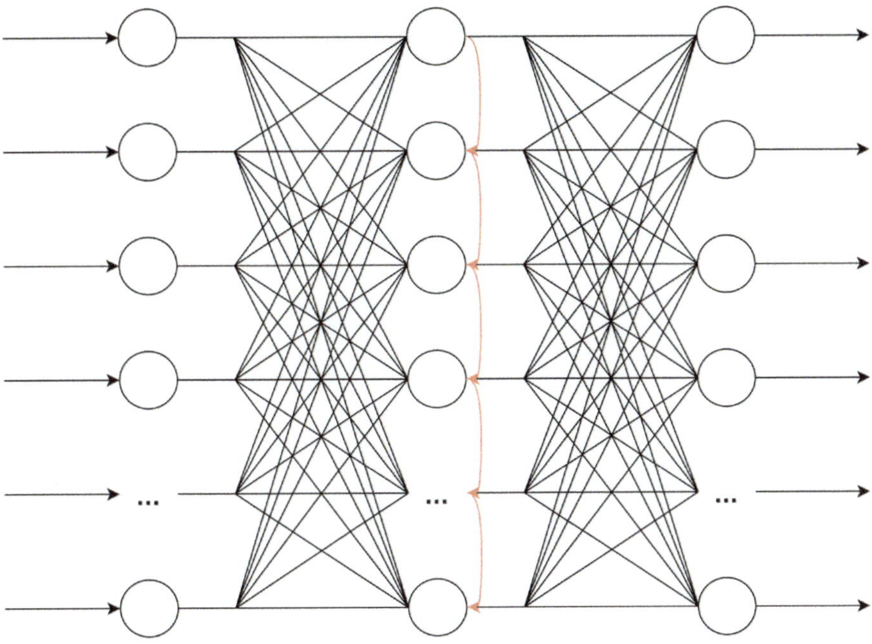

Figure 9.1 Unsupervised SGTM neural-like structure black and white figure.

- The initial value of the transformation step number is setup $P = 1$.
- The initial baseline of the implementation matrix is selected $X_{b,j}^{(P)}$, where $j = \overline{1, n}$, for which the sum of the squares of the elements is maximal. This line corresponds to the vector with the largest norm.
- The coefficient $C_i^{(P)}$, is calculated according to the formula (Berezsky et al., 2023) for all $i = \overline{1, N}$

$$C_i^{(P)} = \frac{\sum_{j=1}^{n} \left(X_{i,j}^{(P)} \bullet X_{b,j}^{(P)} \right)}{\sum_{j=1}^{n} \left(X_{b,j}^{(P)} \right)^2}. \tag{9.1}$$

- The matrix of realizations is transposed, forming a matrix $\left\| X_{j,i}^{T(P)} \right\|$. Further, calculations are performed on its elements according to Eq. (9.2) for all of $j = \overline{1, n}$. As a result, the baseline elements are modified:

$$X_{b,j}^{(P)} = \frac{\sum_{i=1}^{n} \left(X_{j,i}^{T(P)} \bullet C_i^{(P)} \right)}{\sum_{i=1}^{n} \left(C_i^{(P)} \right)^2}. \tag{9.2}$$

- From each row of the training matrix, the elements of the base row are subtracted, multiplied by the coefficient calculated according to Eq. (9.3) for all $i = \overline{1,N}$, and $j = \overline{1,n}$.

$$X_{i,j}^{(P+1)} = X_{i,j}^{(P)} - C_i^{(P)} \bullet X_{b,j}^{(P)}. \tag{9.3}$$

- We proceed to the next step of transformations, performing $P = P + 1$, and transitioning to item 2.
- When the P, reaches the value n, transformations are complete. If the columns of the matrix of implementations do not form linear dependencies, then the resulting matrix will contain only zero elements

$$X_{i,j}^{(P+1)} = \|0\|. \tag{9.4}$$

- The inverse transformation is performed according to the following formula:

$$X_{i,j}^{(P)} = X_{i,j}^{(P+1)} + C_i^{(P)} \bullet X_{b,j}^{(P)}. \tag{9.5}$$

The result of the transformations carried out during SGTM training is a set of vectors $X_{b,j}^{(P)}$, and coefficients $C_i^{(P)}$, for all $P = 1, P_{max}$, $i = 1, N$, $j = 1, n$. These vectors and coefficients determine the parameters of the unsupervised SGTM ANN. Additionally, there is an inverse transformation operation that can be used to restore the original data to these parameters (Tkachenko, 2022; Tkachenko and Izonin, 2019).

In supervised mode, which is also available for SGTM, the data vectors are divided into input and output. However, the value of the coefficient $C_i^{(P)}$, required for successive geometric transformations, can be calculated only for the objects of the training sample. In this case, you can use the value of the coefficient $\overline{C_i^{(P)}}$, which is calculated according to Eq. (9.6), where n is the number of input components in the vector, and represents the desired coefficient as an approximation of the dependence (Tkachenko, 2022; Tkachenko and Izonin, 2019):

$$C_i^{(P)} = f\left(\overline{C_i^{(P)}}\right). \tag{9.6}$$

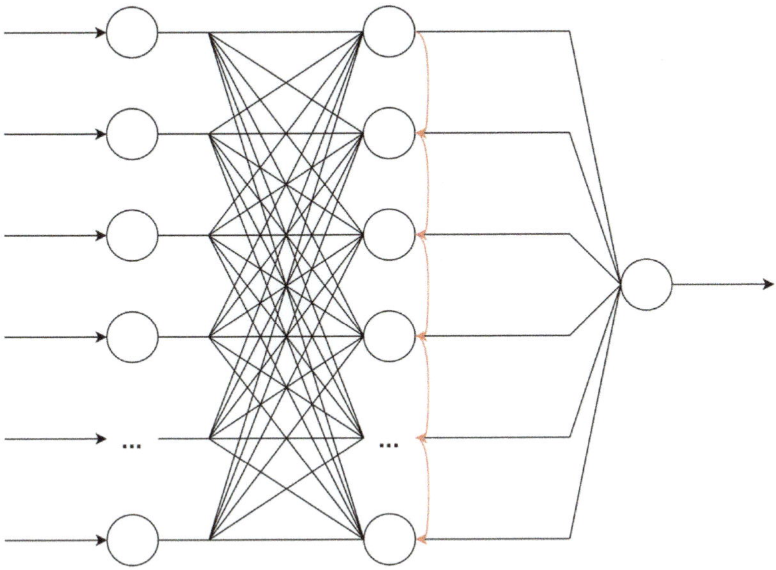

Figure 9.2 Supervised linear SGTM neural-like structure black and white figure.

Fig. 9.2 shows the topology of this SGTM variant, where neural elements are used in the hidden layer, the task of which is to approximate dependence Eq. (9.6).

In the case of the linear variant of ANN, the activation function of neural elements has the following form (Tkachenko, 2022; Tkachenko and Izonin, 2019):

$$C_i^{(P)} = \alpha \times \overline{C_i^{(P)}}, \tag{9.7}$$

where the coefficient α, is calculated by the method of least squares.

Ito decomposition

The Ito decomposition, as a discrete analog of the Voltaire series, is quite often used to approximate complex dependencies (Tkachenko et al., 2018). The quadratic Ito decomposition can be represented as follows:

$$Y(x_1, \ldots, x_n) = a_i + \sum_{i=1}^{n} a_i x_i + \sum_{i=1}^{n} \sum_{j=1}^{n} a_{i,j} x_i x_j. \tag{9.8}$$

In the case of solving a classification task, this decomposition serves as a tool for nonlinear expansion of the task's input data space, which, according to Cover's theorem, will significantly increase the probability of correct classification by the chosen linear machine learning method. In addition, this extension allows us to find new properties of given attributes by modeling the relationships between them. This approach improves classification accuracy. However, the main disadvantage of using such decomposition, especially in the case of a short set of high-dimensional data, is accompanied by a significant increase in the dimensionality of the input data for the neural network. This is why, using high orders of this decomposition can significantly complicate the training of the model, increase the training time, and even provoke an overfitting (Izonin et al., 2019; Tkachenko et al., 2018).

Procedure for the synthesis of linear polynomial

One of the features of SGTM is ensuring the uniqueness of the solution (Izonin et al., 2019; Tkachenko et al., 2018). Taking this into account, it is possible to synthesize the coefficients of a polynomial and replace them with the procedure for using this computational intelligence tool. Procedures for synthesizing the coefficients of a linear polynomial are known (Izonin et al., 2019; Tkachenko et al., 2018). In particular, SGTM ANN with one and with many outputs, the algorithms are described in detail (Izonin et al., 2019; Tkachenko et al., 2018) respectively.

The main steps of the algorithmic implementation of these procedures are the following (Izonin et al., 2019; Tkachenko et al., 2018).

- Training of SGTM ANN on a given set of data with a certain number of independent attributes;
- Formation of the matrix of test signals, as a single diagonal matrix, the number of columns of which is equal to the number of independent attributes of the problem. The number of rows is one more because the first row is added to the single diagonal matrix, the elements of which are zeros to form the free term of the regression or displacement;
- Application of the formed matrix on the pretrained SGTM ANN for the synthesis of the temporary vector of coefficients;
- Synthesis of polynomial terms according to the formula:

$$\forall t \in [1, n+1]: \theta_t = \begin{cases} \alpha_1, t = 1 \\ \alpha_t - \alpha_1, \textit{in other cases} \end{cases}. \tag{9.9}$$

Thus, a transition from a neural network to a polynomial form of application of this AI tool is possible. This will reduce the complexity of using the model, increase the level of interpretability of the results of its work, and, accordingly, increase the level of confidence in the use of this AI tool in practice.

Hybrid cascade-based prediction model and its algorithmic implementations

This subchapter describes the basic provisions of the developed hybrid cascade-based classification method and its three different implementations for improving the efficiency and interpretability of the medical diagnostics results.

Design of the hybrid cascade-based ensemble learning model with improved performance and interpretability

The basis of the developed classification method of short datasets using the hybrid cascade-based ensemble is the consistent use of the naive Bayesian classifier and the SGTM neural-like structure. The main idea of the method is to linearize the response surface, which is fed to the second level of the cascade by taking into account the output signals (as additional independent features) obtained by the algorithm from the first level of the cascade (Izonin and Tkachenko, 2022).

As already mentioned above, the naïve Bayesian classifier allows two variants of the formation of output signals: a label belonging to a specific class of the task and a set of probabilities belonging to each of the defined classes of the task. Since the second option is basic for the formation of a label belonging to a specific observation to a certain class, this option is chosen by us. Such an approach will provide many significant advantages, in particular.

- The possibility of weighing a set of probabilities by neural network tools at the second level of the cascade, which should increase the accuracy of classification;
- The possibility of a significant reduction of independent attributes (for some algorithmic implementations of the method), the number of which in the case of medical diagnostic tasks can be very large;
- The possibility of reducing the duration of the method's training procedure by reducing the space of independent attributes in the data set for the second level of the cascade.

However, the main advantage of the combined use of both of the above methods is the possibility of interpreting the results. This is due to the possibility of switching to the polynomial form of the method, which will be more understandable for medical professionals and workers. Implementation of this approach is possible due to the high interpretive properties of both methods. In particular, the SGTM neural-like structure allows for synthesizing the coefficients of a polynomial, which will provide the possibility of presenting the method results in a nonneural network implementation. The third level of the developed cascade structure will be used for this.

So, the main implementation steps of the hybrid cascade-based classification method using the naïve Bayesian classifier and the SGTM neural-like structures will be the following.

1. Preliminary data processing and formation of training and test datasets.
2. Application of a naive Bayesian classifier to obtain a set of probabilities of belonging to each of the classes of the task.
3. A new data set formation using the above probabilities (variations are possible here, which will be discussed further).
4. Training the SGTM neural-like structure on the newly formed dataset.
5. Synthesis of polynomial coefficients from a pretrained on the previous step, single-output SGTM neural-like structure.

As a result, the proposed method:

- It will increase the efficiency of the intellectual analysis of short datasets (increasing the accuracy of work, increasing the generalization properties of the method as a whole, and reducing the duration of the training procedure for some algorithmic implementations of the method);
- It will increase the result interpretability level (possibility of applying the method in polynomial form, significant reduction of the number of terms of the polynomial for some implementations of the method);
- It will increase the level of understanding by medical specialists as the researched method comes to one or another conclusion, which will increase the level of confidence in its practical use.

Consider several algorithmic implementations of the method. The purpose of their creation is to improve the efficiency of working with high-dimensional or low-dimensional datasets for solving binary or multiclass classification tasks in the case of short datasets specifically in the field of medical diagnostics.

First algorithmic implementation of the proposed method (Algorithm 1)

The first algorithmic implementation of the developed classification method based on a hybrid cascade of two machine learning algorithms is quite simple and intuitive.

The purpose of its development is to increase the accuracy and speed of analysis of high-dimensional short datasets for solving the multiclass data classification task.

The main steps of the Algorithm 1 training procedure.

1. Application of training and test samples to form a set of probabilities of belonging to each of the classes of the task using a naive Bayesian classifier
2. Formation of new datasets by replacing the initial independent attributes with sets of the above probabilities (for training and testing samples). Accordingly, the number of columns of new datasets will be determined by the number of task classes.
3. Training the linear SGTM neural-like structure
4. Synthesis of coefficients of a linear polynomial, according to the algorithm described in the previous subchapter.

The structural and functional scheme of such a hybrid cascade (Algorithm 1) is shown in Fig. 9.3.

The proposed algorithmic implementation of the developed method (Algorithm 1) will show high efficiency in the case of analyzing a high-dimensional dataset for the multiclass data classification task. In theory, a larger number of classes will ensure higher accuracy of Algorithm 1. This will occur due to the possibility of weighting the results of the naive Baye classifier using neural network tools. This very fact, which replaces the procedure of finding the highest probability among the received ones, will in itself ensure an increase in the classification accuracy of medical data.

Second algorithmic implementation of the proposed method (Algorithm 2)

The second algorithmic implementation of the developed classification method based on a hybrid cascade of two machine learning algorithms will solve a different task. The purpose of its development is to increase the accuracy of the analysis of high-dimensional datasets for solving the binary or multiclass data classification task (with a small number of classes defined by the task).

The main steps of the Algorithm 2 training procedure.

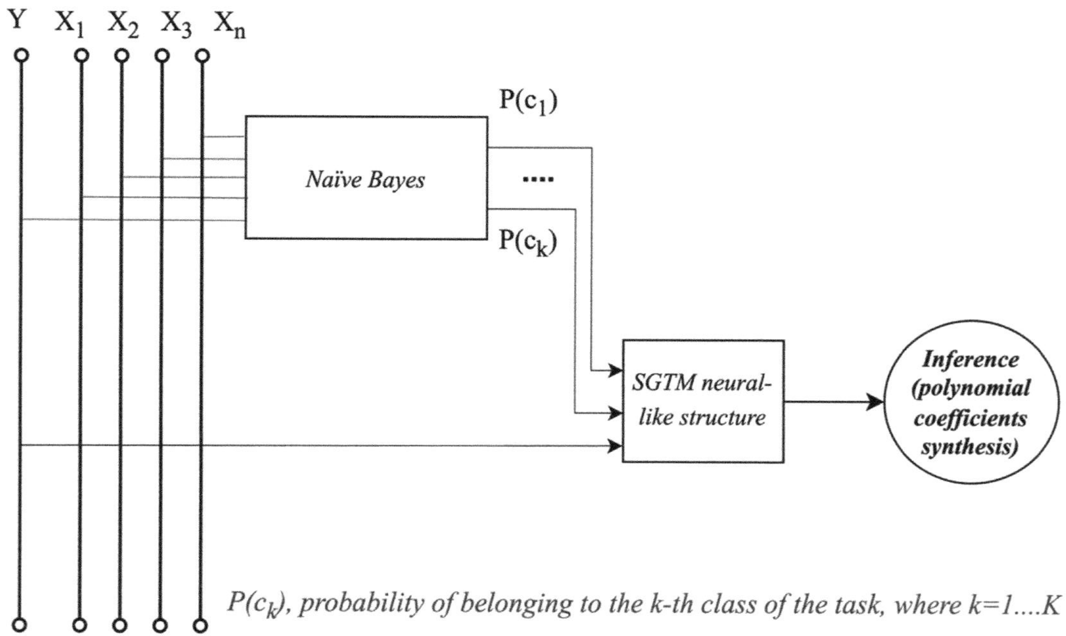

Figure 9.3 Structural and functional scheme of the first algorithmic implementation of the hybrid cascade classifier (Algorithm 1) black and white figure.

1. Application of training and test samples to form a set of probabilities of belonging to each of the classes of the task using a naive Bayesian classifier
2. Formation of new datasets by replacing the initial independent attributes with sets of the above probabilities (for training and testing samples) and their nonlinear expansion using the Ito decomposition of various degrees. Accordingly, the number of columns of the new datasets will be determined by the specified number of task classes.
3. Training linear SGTM neural-like structure
4. Synthesis of the coefficients of the polynomial, in this case, the Ito decomposition of the order specified by the user, according to the algorithm described in the previous subsection.

The structural and functional scheme of such a hybrid cascade (Algorithm 2) is shown in Fig. 9.4.

The proposed algorithmic implementation of the developed method (Algorithm 1) will show high efficiency in the case of analyzing a high-dimensional dataset for tasks with a small number of classes. The nonlinear expansion of the inputs of the SGTM neural-like structure by the Ito decomposition will demonstrate a significant increase in the accuracy of its operation. This is explained by the high approximation properties of this design. However, in the case of a large number of task classes and the use of a large order of the Ito decomposition, the generalization properties of the method may decrease, especially in the case of analyzing a short dataset. This is explained by a significant increase in the independent attributes number of the new dataset, which will be processed by a linear neural-like structure. In this case, it is worth using the first developed algorithm.

Third algorithmic implementation of the proposed method (Algorithm 3)

The third algorithmic implementation of the developed classification method based on a hybrid cascade of two machine learning algorithms solves another task.

The purpose of its development is to increase the accuracy of the analysis of low-dimensional datasets when solving binary or multiclass data classification tasks.

The main steps of the Algorithm 3 training procedure.

1. Application of training and test samples to form a set of probabilities of belonging to each of the classes of the task using a naive Bayesian classifier

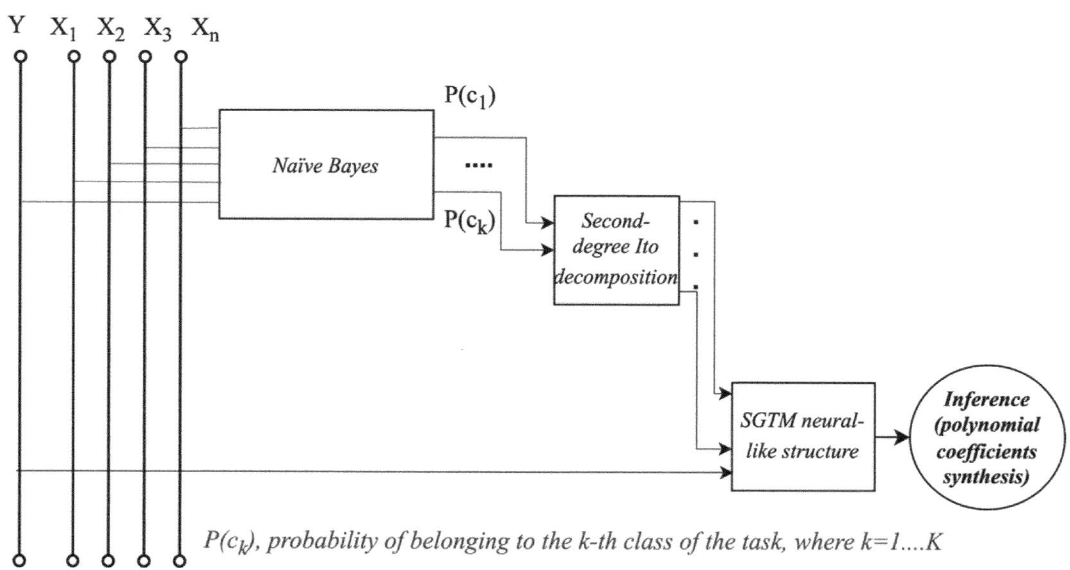

Figure 9.4 Structural and functional scheme of the second algorithmic implementation of the hybrid cascade classifier (Algorithm 2) black and white figure.

2. Formation of new datasets by adding to the initial independent attributes a set of the above probabilities. Accordingly, the number of columns of new sets will increase by the available number of classes defined by a specific task.
3. Training the linear SGTM neural-like structure.
4. Synthesis of coefficients of a linear polynomial, according to the algorithm described in the previous subsection.

The structural and functional scheme of such a hybrid cascade (Algorithm 3) is shown in Fig. 9.5.

The proposed algorithmic implementation of the developed method (Algorithm 3) will show high accuracy in the case of analyzing a low-dimensional dataset for tasks with a large or small number of classes. This is explained by the linearization principle of the response surface due to the consideration of additional independent attributes in the extended dataset. In addition, since these attributes will be a set of probabilities belonging to each of the task's classes, the possibility of their weighting by a neural network tool will provide an additional increase in the accuracy of the developed method. The main disadvantage of this design will be a significant increase in the number of independent attributes, especially in the case of resolving the task with a large number of classes. This will increase the duration of the method training procedure, however, in the case of analysis of short datasets, this disadvantage can be eliminated.

Datasets description and their preprocessing

This subchapter describes the datasets used to model the performance of the designed method, their preprocessing procedures, and the approach to balancing short medical datasets used in this chapter.

Dataset 1

The first dataset is taken from an open repository (Uci, 2023). It contains information on critically ill patients, data on which were collected at various medical centers in the United States. The purpose of intellectual analysis in our case is to predict the survival of critically ill patients 180 days after hospitalization.

The dataset required the performance of several preliminary processing tasks, which were done in this work before intellectual analysis (binary classification task).

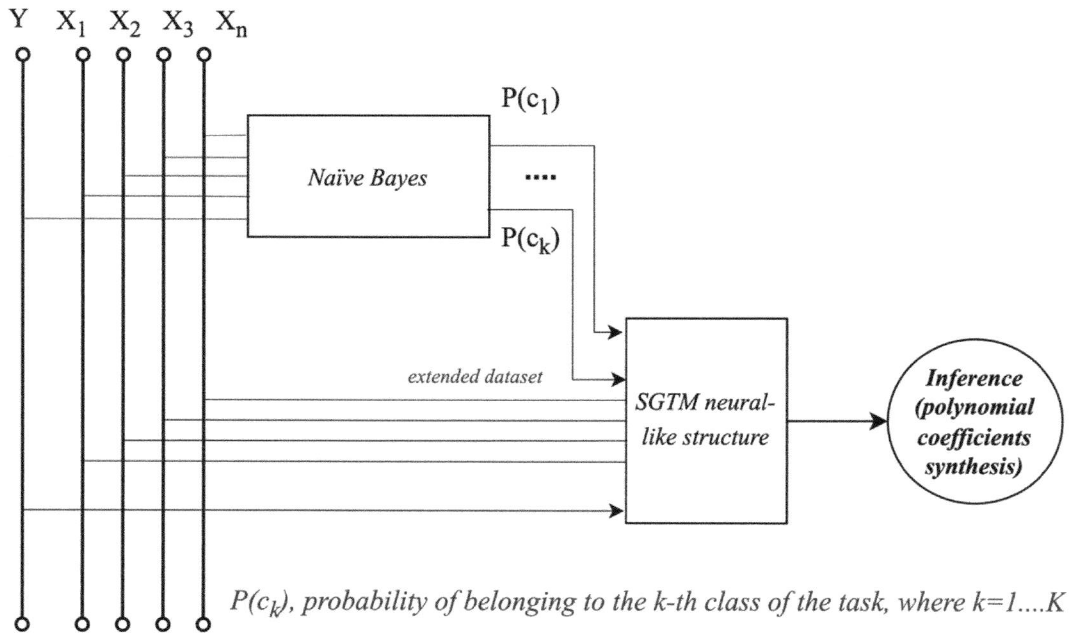

Figure 9.5 Structural and functional scheme of the third algorithmic implementation of the hybrid cascade classifier (Algorithm 3) black and white figure.

Data preprocessing consisted of:

- NaN values were replaced with "n/a" for categorical columns;
- NaN values were replaced with average column values for numerical columns;
- The One-hot encoding was used for encoding categorical columns;

As a result, preprocessed data consisted of 9105 samples, characterized by 67 attributes.

Dataset 2

The second dataset is also taken from an open repository (Cirrhosis Patient Survival Prediction, 2023). It contains information about patients with liver cirrhosis. The purpose of intellectual analysis in this case is to predict the survival of patients after liver transplantation. This data set also required several preprocessing tasks, which were done in this work before intelligent analysis (multiclass classification task).

Data preprocessing consisted of:

- NaN values were replaced with "N/a" for categorical columns;
- Columns 'Cholesterol', 'Copper', 'Triglycerides', and 'Platelets' were converted to int datatype;
- "NaNN" values were replaced with average column values for numerical columns 'Cholesterol', 'Copper', 'Triglycerides',
- NaN values were replaced with average column values for numerical columns;
- The One-hot encoding was used for encoding categorical columns;
- The target column was encoded using mapping; ('D': 0, 'C': 1, 'CL': 2)

As a result, preprocessed data consisted of 418 samples, characterized by 28 attributes.

Dataset 3

A third dataset is also listed in an open repository (Bone Marrow Transplant, 2023). It contains information about children with various diseases who need bone marrow transplants. The purpose of intellectual analysis in this case is to predict the survival of patients after bone marrow transplantation. This data set also required some preprocessing tasks, which were done in this work before the intelligent analysis (binary classification task).

Data preprocessing consisted of:

- The 'Disease' column was dropped;
- NaN values were replaced with average column values for numerical columns;

As a result, preprocessed data consisted of 187 samples, characterized by 35 attributes.

Balancing techniques for small data analysis

A high-quality, representative, balanced dataset is very important for building a reliable machine learning-based classification model (Jangra et al., 2023). One of the important features of medical data is that it is usually not balanced. This is explained by the specificity of the field of medicine, an unbalanced data set is difficult to obtain (Tolstyak et al., 2023; Tolstyak and Havryliuk, 2022).

Most of the existing classification methods usually show unsatisfactory results when processing representatives of a smaller class in conditions of highly unbalanced data. This is explained by the fact that the main focus of such methods is on optimizing the overall accuracy of the model. That is, the setting of synaptic weights of artificial neural networks takes place by the received training sample while ignoring the distribution of representatives of each class. First of all, this leads to the fact that the model will have a high accuracy rate due to correct predictions for a larger class. However, errors in the classification of the smaller class, which are especially difficult to collect in medicine, are more important and can lead to serious consequences. In addition, such models will not be able to generalize correctly to new data, since they will be trained on specific features of a larger class.

This is why balancing the studied dataset will ensure the possibility of obtaining objective and reliable results of a specific machine learning model, as well as correctly taking into account all classes of the task.

Among the main classes of balancing methods, two main ones can be distinguished.

- balancing at the data level (adding instances to a smaller class, removing some observations from a larger class, or using them together)
- balancing at the algorithmic level (class weight adjustment - classification error prices, setting the decision threshold, etc.).

Since this work is devoted to the task of intelligent analysis of short datasets, the application of methods for removing observations from a larger class is not acceptable. Balancing methods at the algorithmic level also have some disadvantages. In particular, they require an expert to set the weights of the task classes. This imposes many limitations on their practical application to rare medical problems. That is why, in this chapter, special attention is devoted to SMOTE (Synthetic Minority Over-sampling

Technique) - one of the fast and practical algorithms from the class of balancing at the data level.

The main steps of the algorithmic implementation of the synthesis of new samples of the minor class according to the SMOTE method are the following (Soltanzadeh and Hashemzadeh, 2021).

1. to search for differences between the current instance and its nearest neighbor;
2. to synthesize random deviation by multiplying the found differences by a random number in the range [0,1]
3. to synthesis of a new instance by adding a random deviation to the current instance
4. to perform the above actions with other neighbors of the current instance until obtaining the required number of new representatives of the minor class.

It is this approach that is used in this paper to balance all the studied datasets.

Modeling and results

The modeling of the developed method and its three algorithmic implementations took place on a computer with the following parameters: processor core i3 (10th gen), the number of cores: 2, and the amount of RAM: 8 GB. The effectiveness of his work was evaluated using three criteria.

1. The accuracy scores are based on six different indicators;
2. The duration of the training procedure, in seconds;
3. An interpretability, based on the polynomial representation of the operation of the hybrid cascade in the application mode.

The accuracy scores of the three algorithmic implementations of the developed cascade method were evaluated based on six indicators.

- Total accuracy score;
- Precision score;
- Recall score;
- F1-score;
- Cohen's Kappa score (Kappa);
- Matthews Correlation Coefficient score (MCC).

A detailed description of these metrics, together with their mathematical representation, is given in (Chicco et al., 2021).

All three real-world datasets were randomly divided into training (80%) and test (20%) samples. Next, the data were normalized using the maximum element normalization method in each data column.

Data balancing results

Since all three datasets are unbalanced, the paper investigated the application of the balancing technique to the accuracy of the operation of all three modifications. The authors used SMOTE to synthesize instances of a smaller class or classes (depending on the task) to balance the number of representatives of each task's class. Next, the simulation of machine learning methods was carried out using the new datasets obtained after balancing.

Table 9.1 summarizes the results of the accuracy of the application of both studied classifiers when solving medical diagnosis task using dataset 1.

As can be seen from Table 9.1, the use of SMOTE significantly increased the accuracy of the SGTM neural-like structure (by 12% according to the F1-score in the application mode), but did not change, or worsen the results of the Naïve Bayes classifier.

Table 9.2 summarizes the results of the accuracy of the application of both studied classifiers when solving the medical diagnosis task using dataset 2.

As can be seen from Table 9.2, the use of SMOTE greatly increased the accuracy of the SGTM neural-like structure (based on all investigated indicators), but significantly worsened the results of the Naïve Bayes classifier.

Table 9.3 summarizes the results of the accuracy of the application of both studied classifiers when solving medical diagnosis tasks using dataset 3.

As can be seen from Table 9.3, the use of SMOTE slightly increased the accuracy of the SGTM neural-like structure (based on all investigated indicators), but significantly increased the results of the Naïve Bayes classifier. This is due to the significant imbalance of the original data set.

Despite such different results, for further research, it was decided to use the balancing technique to build more reliable methods of intellectual analysis of such data.

Implementation results of the Algorithm 1

The first algorithmic implementation of the developed cascade method (Algorithm 1) was used to solve the task of predicting the survival rate of critically ill patients 180 days after hospitalization using dataset 1. This is a binary classification task in the case of an unbalanced dataset. Because of

Table 9.1 Results of data balancing for the Dataset 1.

Method/ performance indicator	Train F1 score	Test F1 score	Train Kappa	Test Kappa	Train MCC	Test MCC
Original data						
Naïve Bayes	0.70	0.57	0.38	0.01	0.42	0.05
SGTM neural-like structure	0.77	0.63	0.47	0.17	0.47	0.20
Data balanced using SMOTE techniques						
Naïve Bayes	0.73	0.56	0.46	0.00	0.46	0.03
SGTM neural-like structure	0.76	0.75	0.53	0.40	0.53	0.40

Table 9.2 Results of data balancing for the Dataset 2.

Method/ performance indicator	Train F1 score	Test F1 score	Train Kappa	Test Kappa	Train MCC	Test MCC
Original data						
Naïve Bayes	0.63	0.53	0.34	0.25	0.39	0.28
SGTM neural-like structure	0.68	0.59	0.44	0.19	0.47	0.19
Data balanced using SMOTE techniques						
Naïve Bayes	0.69	0.42	0.58	0.08	0.54	0.16
SGTM neural-like structure	0.37	0.54	0.32	0.32	0.15	0.14

Table 9.3 Results of data balancing for the Dataset 3.

Method/ performance indicator	Train F1 score	Test F1 score	Train Kappa	Test Kappa	Train MCC	Test MCC
Original data						
Naïve Bayes	0.64	0.52	0.31	0.22	0.43	0.35
SGTM neural-like structure	0.95	0.88	0.90	0.75	0.90	0.75
Data balanced using SMOTE techniques						
Naïve Bayes	0.73	0.56	0.49	0.26	0.56	0.39
SGTM neural-like structure	0.96	0.88	0.92	0.75	0.92	0.75

this, the authors used several different indicators for evaluating the effectiveness of the method in both the training and application modes of Algorithm 1.

Table 9.4 summarizes the values of all studied metrics in both of the above modes. It should be noted that performance indicators were calculated after the second level of the developed cascade-based classification method.

As can be seen from Table 9.4, the accuracy of Algorithm 1 on the first dataset is satisfactory based on all the studied metrics. In addition, it demonstrates high generalization properties.

The use of the third level of the cascade classifier, according to Algorithm 1, provides the possibility of leveling the second level of the cascade form of the application of the developed method and the transition to the polynomial form of its use. Considering one of the advantages of using SGTM neural-like structures in the structure of the cascade, which was already mentioned in the previous sections, it is possible to obtain a linear polynomial, which will be an analog of the application mode of the method. However, in this case, the results can be easily interpreted by a medical professional.

According to the structure of the algorithmic implementation of the developed hybrid cascade method (Algorithm 1) and taking into account the input data (dataset 1), the authors synthesized the coefficients of a linear polynomial, which can be used instead of performing the second layer of the method. They are summarized in Table 9.5.

The obtained linear polynomial in the case of Algorithm 1 operates with only two independent attributes to solve the stated task. In this way, we managed to reduce the dimensionality of the input data space from 67 attributes to 2. This significantly reduced the duration of the training

Table 9.4 Implementation results of Algorithm 1 using Dataset 1.

Performance indicator/ algorithm's mode	Training mode	Test mode
Total accuracy	0.74	0.73
Precision	0.76	0.76
Recall	0.74	0.73
F1-score	0.74	0.74
Kappa Coena	0.43	0.43
MCC	0.44	0.44

Table 9.5 Linear polynomial coefficients for Algorithm 1.

Term of a linear polynomial	The value of the coefficient for the corresponding term
Displacement	000,000,000
P(1)	037,514,902
P(2)	091,247,915

where: P(1) and P(2) are probabilities of belonging to two problem classes, respectively, obtained by a naive Bayesian classifier.

procedure, taking into account the number of instances in this dataset, while obtaining satisfactory classification accuracy. It should be noted that the obtained linear polynomial was used to check the accuracy of its operation in comparison with the accuracy of the cascade classifier (Algorithm 2). Identical results were obtained.

Implementation results of the Algorithm 2

The second algorithmic implementation of the developed cascade method (Algorithm 2) was used to solve the task of predicting the survival rate of patients with liver cirrhosis after transplantation of this organ. The simulation was performed using dataset 2. This is a multiclass classification task in the case of an unbalanced data set. That is why several different indicators for evaluating the effectiveness of the method in the modes of both training and application of Algorithm 2 were also used here.

Table 9.6 summarizes the values of all studied metrics in both of the above modes. It should be noted that performance indicators were calculated after the second level of the developed cascade classifier.

As can be seen from Table 9.6, the accuracy of Algorithm 2 on the second dataset is also satisfactory based on all the studied metrics.

Table 9.6 Implementation results of Algorithm 2 using Dataset 2.

Performance indicator/ algorithm's mode	Training mode	Test mode
Total accuracy	0.73	0.74
Precision	0.69	0.71
Recall	0.73	0.74
F1-score	0.70	0.71
Kappa Coena	0.48	0.48
MCC	0.48	0.46

According to the structure of the algorithmic implementation of the developed hybrid cascade classifier (Algorithm 2) and taking into account the input data (dataset 2), the authors synthesized coefficients of the second-order Ito decomposition, which can be used instead of performing the second layer of the method. They are summarized in Table 9.7 second-degree.

The obtained second-degree Ito decomposition contains only nine independent attributes for solving the stated task. In this way, we managed to reduce the size of the dataset for analysis from 28 attributes to 9, which accordingly reduced the duration of the method training procedure as a whole. It should be noted that the obtained second-order Ito decomposition was used to check the accuracy of its operation in comparison with the accuracy of the cascade classifier (Algorithm 2). Identical results were obtained.

Implementation results of the Algorithm 3

The third algorithmic implementation of the developed cascade classifier (Algorithm 3) was used to solve the task of predicting the survival rate of children with various serious diseases after their bone marrow transplantation. Experimental studies were conducted using dataset 3. This is a binary classification task in the case of an unbalanced dataset. In this case, the authors also used several different indicators for evaluating the effectiveness of the method in both the training and application modes of Algorithm 3.

Table 9.7 Coefficients of the second-degree Ito decomposition for Algorithm 2.

A member of the Ito decomposition of the second-degree	The value of the coefficient for the corresponding term
Displacement	00,000,000,000
P(1)	07,945,450,200
P(2)	09,546,040,700
P(3)	07,620,143,200
P(1)*P(2)	02,054,551,100
P(1)*P(3)	02,508,477,200
P(2)*P(3)	04,434,473,700
P(1)*P(1)	00,453,962,700
P(2)*P(2)	02,833,820,900
P(3)*P(3)	02,379,853,000

where: P(1), P(2), and P(3) are the probabilities of belonging to the three task classes, respectively, obtained by a naive Bayesian classifier.

Table 9.8 Implementation results of Algorithm 3 using Dataset 3.

Performance indicator/ algorithm's mode	Training mode	Test mode
Total accuracy	0.97	0.89
Precision	0.97	0.95
Recall	0.97	0.89
F1-score	0.97	0.92
Kappa Coena	0.93	0.82
MCC	0.93	0.80

Table 9.8 summarizes the values of all studied metrics in both of the above modes. It should be noted that performance indicators were calculated after the second layer of the developed cascade classifier.

As can be seen from Table 9.8, the accuracy of Algorithm 3 on the third dataset is quite high based on all investigated metrics, considering the size of the dataset. In addition, the cascade classifier does not exhibit overfitting, which is typical when analyzing short datasets.

According to the structure of the algorithmic implementation of the developed hybrid cascade method (Algorithm 3) and taking into account the input data (data et 3), the authors synthesized

coefficients of a linear polynomial, which can be used instead of performing the second layer of the method. They are summarized in Table 9.9.

In this case, the linear polynomial contains additional coefficients about 35 initial independent attributes, which correspond to the set of probability of belonging to one of the two classes of the task. However, increasing the size of the input data space from 35 to 37 attributes did not significantly affect the duration of the training procedure considering that the dataset is quite short. It should be noted that the obtained linear polynomial was used to check the accuracy of its operation in comparison with the accuracy of the cascade classifier (Algorithm 3). Identical results were obtained.

Comparison and discussion

This subchapter presents the results of comparing three developed algorithmic implementations of the method with parent machine learning algorithms that form the method according to three different criteria. Let's consider them in more detail.

Table 9.9 Linear polynomial coefficients for Algorithm 3.

Term of a linear polynomial	The value of the coefficient for the corresponding term	Term of a linear polynomial	The value of the coefficient for the corresponding term	Term of a linear polynomial	The value of the coefficient for the corresponding term
Displacement	−0,00,643,176	X_{13}	134,813,393	X_{26}	123,673,351
X_1	088,166,894	X_{14}	019,081,098	X_{27}	027,966,786
X_2	070,597,988	X_{15}	096,315,076	X_{28}	105,357,264
X_3	121,825,382	X_{16}	−0,44,588,487	X_{29}	123,242,002
X_4	079,330,708	X_{17}	102,161,335	X_{30}	016,148,671
X_5	081,718,098	X_{18}	03,996,255	X_{31}	084,006,435
X_6	006,072,325	X_{19}	019,468,503	X_{32}	051,936,306
X_7	047,396,978	X_{20}	103,049,583	X_{33}	023,605,242
X_8	062,287,312	X_{21}	077,866,586	X_{34}	−0,22,070,569
X_9	007,548,170	X_{22}	−1,38,731,853	X_{35}	−0,51,684,824
X_{10}	079,955,098	X_{23}	062,466,265	P(1)	113,185,872
X_{11}	068,193,858	X_{24}	060,520,983	P(2)	069,300,241
X_{12}	006,746,652	X_{25}	032,653,249		

where: X_1 … X_{35} are the initial inputs of the task (initial independent variables); P(1) and P(2) are the probabilities of belonging to two task classes, respectively, obtained by a naive Bayesian classifier.

Comparison of the accuracy of all investigated methods

To compare the accuracy of each of the developed algorithmic implementations of the designed cascade classifier, the authors use two machine learning algorithms that are the basis of each of the algorithmic implementations of the cascade classifier.

Figs. 9.6–9.8 show the histograms of the comparison of the accuracy of all the studied methods using different performance indicators for three algorithmic implementations of the method (Algorithm 1, Algorithm 2, and Algorithm 3, respectively).

Summarizing the comparison results of three algorithmic implementations of the method, the following can be stated.

- Basic machine learning methods demonstrate unsatisfactory classification accuracy when solving all three tasks;
- Each of the developed algorithms provided a significant increase in accuracy compared with the basic machine learning methods;
- The developed algorithms are not prone to overfitting and demonstrate high generalization properties;
- Algorithm 1 reduced the space of input data of the task from 67 to 2 attributes and increased the classification accuracy by 26% (based on Cohen's Kappa - Kappa). When analyzing the first dataset, the baseline algorithms show significant overfitting, while Algorithm 1 eliminates this problem;

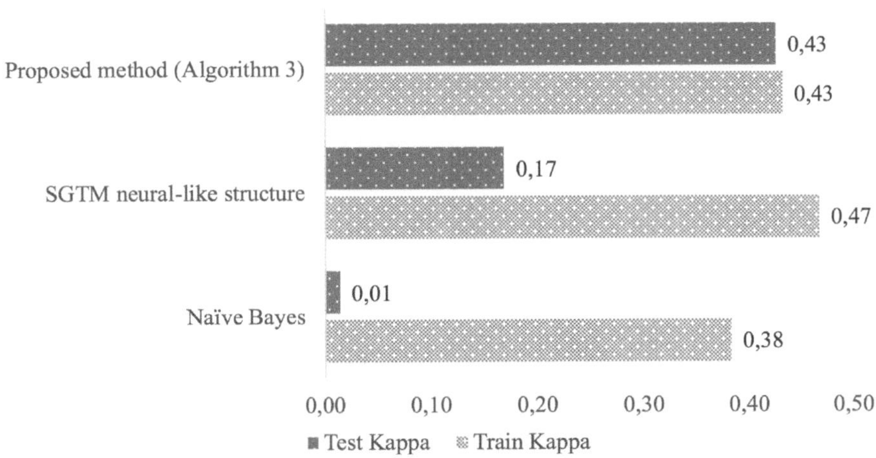

Figure 9.6 Accuracy comparison of the proposed method (Algorithm 1) with the accuracy of parental ML-based methods that form it based on the use of Cohen's Kappa score (using Dataset 1) Grayscale figure.

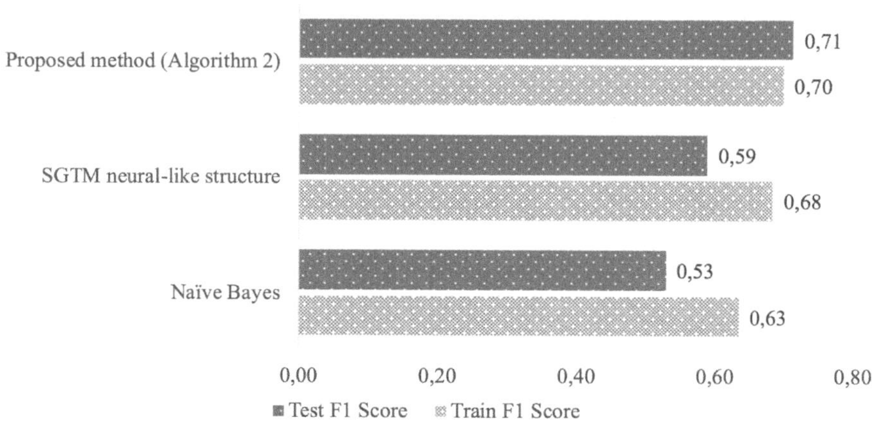

Figure 9.7 Accuracy comparison of the proposed method (Algorithm 2) with the accuracy of parental ML-based methods that form it based on the use of F1-score (using Dataset 2) Grayscale figure.

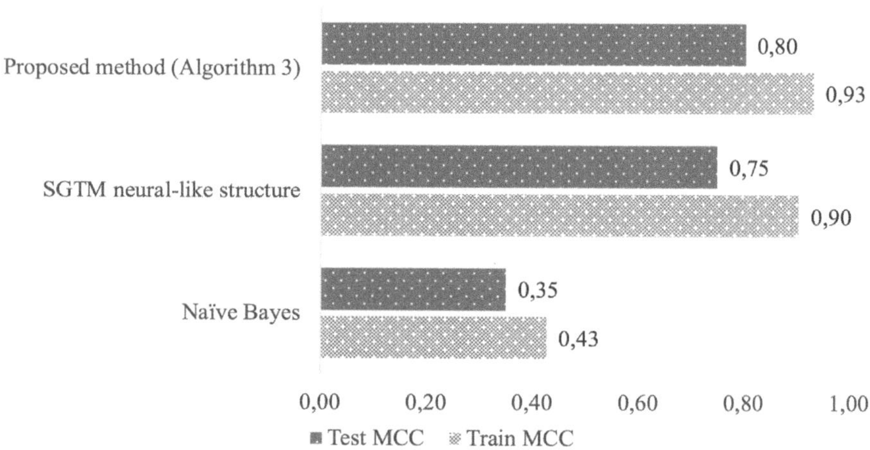

Figure 9.8 Accuracy comparison of the proposed method (Algorithm 3) with the accuracy of parental ML-based methods that form it based on the use of Matthews correlation coefficient (using Dataset 3) Grayscale figure.

- Algorithm 2 increased classification accuracy by 12% (based on F1-score). In addition, it provided a reduction in the task's input data space from 28 to 9 attributes. Basic machine learning algorithms when analyzing dataset 2 demonstrate overfitting;
- Algorithm 3 increased the classification accuracy by 5% (Matthews Correlation Coefficient - MCC) with significantly higher generalization properties compared with the basic machine learning methods.

Comparison of the training duration for all investigated methods

An important indicator to evaluate the effectiveness of cascaded ensembles is the duration of the training procedure implementation. Usually, cascade ensembles use a highly developed hierarchical structure to ensure high classification accuracy, each level of which significantly complicates the training procedure. Accordingly, the duration of such procedures can be significant even in the case of analysis, given the fact that such ensembles cannot be parallelized (Mochurad and Shchur, 2021). In addition, one of the algorithms uses the Ito decomposition, which with large degrees, can significantly expand the already large sets of medical data (which contain many independent attributes) that affect the possibility of forming a correct diagnosis.

That is why this subchapter compares the duration of the training procedure of all three algorithmic implementations of the method with the basic AI tools, which are the basis of the developed method.

Figs. 9.9—9.11 summarize the results for three algorithmic implementations of the developed method, respectively.

Figs. 9.9—9.11 show the histograms of the comparison of the accuracy of all the studied methods using different performance indicators for three algorithmic implementations of the method (Algorithm 1, Algorithm 2, and Algorithm 3, respectively).

Summarizing the obtained results, it is possible to state the following.

- the duration of the training procedure of all investigated methods is satisfactory, which is explained by the small amount of data for analysis;

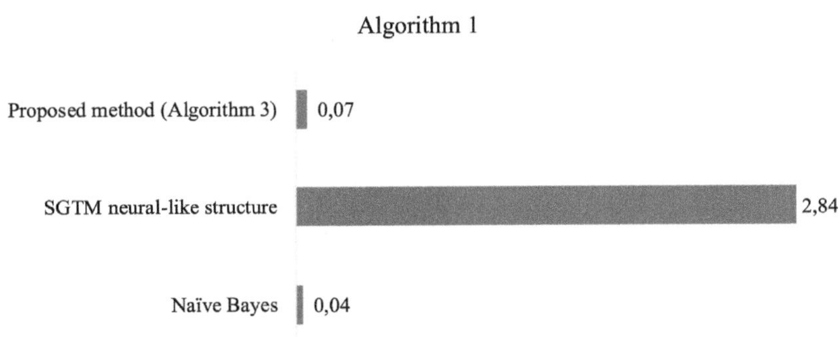

Figure 9.9 Comparison of the training time (seconds) for the proposed method (Algorithm 1) and parental ML-based methods that form it (using Dataset 1) Grayscale figure.

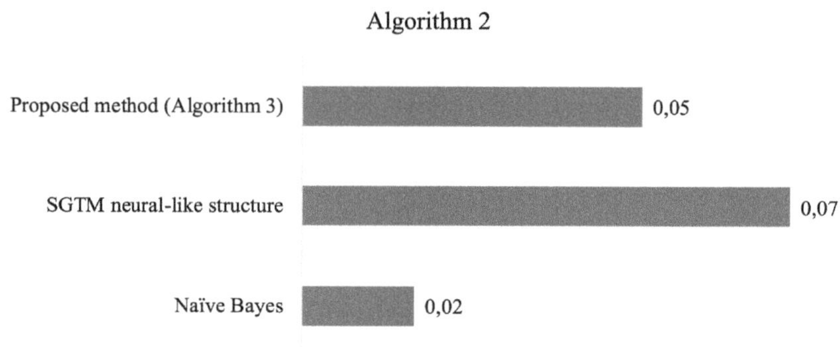

Figure 9.10 Comparison of the training time (seconds) for the proposed method (Algorithm 2) and parental ML-based methods that form it (using Dataset 2) Grayscale figure.

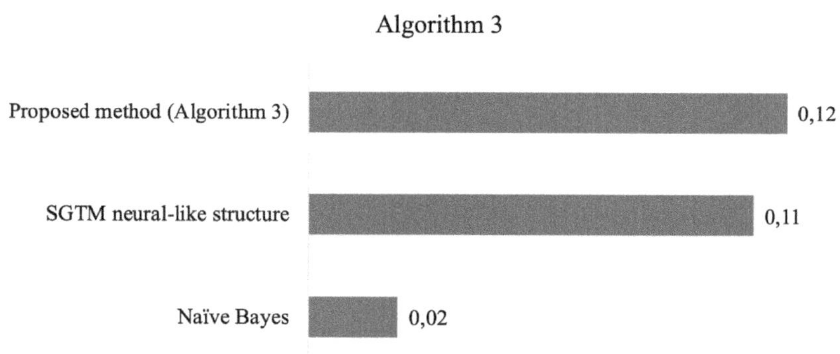

Figure 9.11 Comparison of the training time (seconds) for the proposed method (Algorithm 3) and parental ML-based methods that form it (using Dataset 3) Grayscale figure.

- the shortest training time is shown by the naive Bayesian classifier, which is explained by the peculiarities of its operation and implementation;
- a linear SGTM neural-like structure demonstrates the highest training time, which is explained by the high dimensionality of all three datasets;
- the developed cascade-based classifier demonstrates a significantly shorter duration of the training procedure compared with the SGTM neural-like structure during the processing of the first and second datasets. This is explained by the significantly reduced dimensionality of the dataset, which is processed at the second level of the cascade for Algorithm 1 and Algorithm 2.

- Algorithm 4 requires the most time to implement the training procedure, which is explained by the increase in the space of input data, in particular from 35 to 37 in the case of solving the stated task using dataset 2.

Comparison of the interpretability of all investigated methods

In addition to the accuracy and speed of operation of the selected artificial intelligence tools, the interpretation of the results of their work is one of the important indicators of the use of such tools in practice. First of all, this is explained by the level of trust or distrust in the particular method by a medical specialist. In the general case, the medical specialist wants to understand how the selected method reached one or another conclusion. A high level of results interpretation of the selected artificial intelligence tool increases the level of trust and, accordingly, the practical application level of the method for solving various applied problems, in particular in various fields of medical diagnostics.

Today, there are many explainable AI models and methods. However, one of the simplest and at the same time understandable is the possibility of presenting the results in polynomial form. In our case, we can ensure this due to the possibility of transition from the neural network application of the developed model (which is essentially a black box for a medical specialist) to the polynomial form of the model application, which will be able to achieve the same efficiency results, but will be a clear and understandable in practice. This approach is implemented due to the use of SGTM neural-like structures in the structure of the developed method.

In fact, in the case of using algorithmic implementations of the developed method, including the second level of the cascade, the specialist will simply get the result - i.e., a mark of whether a specific observation belongs to one of the defined classes of the problem. However, the use of the third level of the developed cascade provides the possibility of synthesizing the coefficients of a linear or nonlinear polynomial and a partial transition from the cascading form of the method's application to the polynomial one. In this case, partial transition means that we should in any case obtain a set of probabilities using a naive Bayesian classifier. This will be the first step. In the next step, we will substitute them into a linear polynomial synthesized for a specific algorithmic implementation of the method to obtain the result.

Since the naive Bayesian classifier is one of the simplest explainable AI tools, the process of obtaining a set of probabilities is understandable. Further, their use by each of the developed algorithms will bring its own, quite different level of interpretability of the result. Let's consider each algorithm in more detail.

The high-dimensional dataset (binary classification task) used to model the performance of Algorithm 1, contains 67 independent attributes. Using a naive Bayesian classifier to solve the stated task will provide only two probabilities, which is very clear and easy to use for a medical practitioner. However, as shown above, the accuracy of this algorithm is not satisfactory. The direct application of the SGTM neural-like structure to solve this task will allow us to synthesize a linear polynomial, which can also be a convenient and understandable form of interpretation of the result. However, in addition to the unsatisfactory accuracy of its work, the number of members of such a polynomial will be 67. The interpretation of such a polynomial model as well as its practical application will cause several problems due to the large number of coefficients. The first algorithm of the developed method, which combines the use of a naive Bayesian classifier and SGTM neural-like structures, allows synthesizing a polynomial from only two independent variables (three coefficients, where the first is a free regression member or displacement, and the other two are coefficients near the probabilities of belonging to each of the two classes tasks). Compared with the previous methods, it will be clear, very easy to use, and, as the results showed, effective in terms of accuracy and speed. It is obvious that, in this case, the level of interpretability of the results of Algorithm 1 (3 coefficients) will be significantly higher compared with the level of interpretability of the results of the SGTM neural-like structure work (67 coefficients), and therefore will increase the overall level of confidence in the practical use of the developed cascade classifier.

The second set of data for solving the multiclass classification task (for three classes) was used to model Algorithm 2. In this case, the use of a naive Bayesian classifier in the form of a synthesis of three probabilities belonging to each class of the task will be the most convenient and interpretable form of AI. SGTM neural-like structures, in this case, can synthesize a polynomial with 28 coefficients that can also be interpreted. However, both methods do not provide satisfactory classification accuracy. However, their combined use in the form of Algorithm 2 of the developed cascade classifier makes it possible to synthesize a nonlinear polynomial of the second degree according to the Ito decomposition based on a set of three probabilities. Thus, the resulting polynomial will have only nine members, which will have a higher level of interpretation and will be more convenient for practical use.

Algorithm 3 of the developed cascade classifier uses a third dataset that marks 35 independent attributes and consists of such a combination of two machine learning methods that the resulting linear polynomial will contain 37 terms. This result will demonstrate almost the same level of interpretation as the use of the SGTM neural-like structure (35 polynomial terms), but it shows a much higher accuracy in solving the medical diagnosis task. Due to such a large number of members of the resulting polynomial, this algorithmic implementation of the developed cascade classifier (Algorithm 3) should be used when processing low-dimensional sets of medical data.

In summary, the polynomial form of using the developed method (linear or nonlinear polynomial) will provide many advantages, in particular.

- A significant increase in the confidence level in the use of the developed method, regardless of the selected algorithmic implementation;
- Significant reduction of the independent attributes number (for Algorithms 1 and 2), which will make the polynomial model clearer and easier to use;
- The possibility of evaluating the importance of independent attributes according to the value of the coefficients synthesized by the corresponding algorithm. In this case, a larger coefficient will determine a greater influence of the corresponding independent attribute on the formation of the result, which will allow additional data analysis and interpretation.

Conclusions

This section is devoted to the problem of intelligent data analysis by AI tools in the field of medical diagnostics with the possibility of interpreting such analysis results. The authors developed a new hybrid cascade of two machine learning methods. The proposed ensemble for the classification of short datasets is based on the consistent use of a naive Bayesian classifier and an SGTM neural-like structure. Their joint use ensures an increase in efficiency (accuracy, training time) and the level of interpretability (in polynomial form) of classification results. In particular, the efficiency of the method is increased due to the linearization of the response surface, which is fed to the second level of the cascade by taking into account the output signals (as additional independent features) obtained by the algorithm from the first level of the cascade. Interpretability is provided by the possibility of synthesizing the coefficients of a linear or nonlinear polynomial and thus the transition from the neural network-to the polynomial form of the developed method.

The practical implementation of the developed cascade method is performed in the form of three different algorithms, each designed to solve a certain class of medical diagnostic tasks. Modeling of the method was carried out on three sets of medical data. Performance indicators based on various metrics are presented, and polynomials are synthesized for the transition from neural network to polynomial form of developed method application. A comparison between the method's work and parental machine learning algorithms, which are the basis of its work, is performed. A significant increase in accuracy, a reduction in the duration of the training procedure, and a high level of interpretation of each of the developed algorithmic implementations of the method were demonstrated.

The developed cascade classifier can be used in other application areas, such as (Alsayaydeh et al., 2021; Mochurad et al., 2021; Indha et al., 2020; Semkovych and Shymanskyi, 2023).

Funding

This research was funded by the European Union (through the EURIZONE H2020 project, grand agreement 871072).

References

Alsayaydeh, J.A.J., Aziz, A., Rahman, A.I.A., Salim, S.N.S., Zainon, M., Baharudin, Z.A., Abbasi, M.I., Khang, A.W.Y., 2021. Development of programmable home security using GSM system for early prevention. ARPN J. Eng. Appl. Sci. 16 (1), 88–97, 18196608.

Berezsky, O., Pitsun, O., Liashchynskyi, P., Derysh, B., Batryn, N., 2023. Computational intelligence in medicine. Lect. Data Eng. Communicat. Technol. 149, 488–510. https://doi.org/10.1007/978-3-031-16203-9_28, 23674520.

Berezsky, O., Dubchak, L., Batryn, N., Datsko, T., Berezska, K., Pitsun, O., Batko, Y., 2019. Fuzzy system for breast disease diagnosing based on image analysis. In: CEUR Workshop Proceedings, pp. 69–83, 2488, 16130073.

Bisikalo, O., Kharchenko, V., Kovtun, V., Krak, I., Pavlov, S., 2023. Parameterization of the stochastic model for evaluating variable small data in the shannon entropy basis. Entropy 25 (2). https://doi.org/10.3390/e25020184, 1099–4300.

Bone Marrow Transplant: Children, 2023. UCI Machine Learning Repository n.

Chicco, D., Tötsch, N., Jurman, G., 2021. The Matthews correlation coefficient (Mcc) is more reliable than balanced accuracy, bookmaker informedness, and markedness in two-class confusion matrix evaluation. BioData Min. 14, 1–22. https://doi.org/10.1186/s13040-021-00244-z, 17560381.

Cirrhosis Patient Survival Prediction - UCI Machine Learning Repository n, 2023.

Goyal, L., Dhull, A., Singh, A., Kukreja, S., Singh, K.K., 2022. VGG-COVIDNet: a novel model for COVID detection from X-Ray and CT Scan images. Proc. Comput. Sci. 218, 1926–1935. https://doi.org/10.1016/j.procs.2023.01.169, 18770509.

Hovorushchenko, T., Hnatchuk, Y., Herts, A., Onyshko, O., 2021. Intelligent information technology for supporting the medical decision-making considering the legal basis. In: CEUR Workshop Proceedings, pp. 72–82, 2853, 16130073.

Indha, W.A., Zamzam, N.S., Saptari, A., Alsayaydeh, J.A., Hassim, N.B., 2020. Development of security system using motion sensor powered by RF energy harvesting. In: 2020 IEEE Student Conference on Research and Development, SCOReD 2020. Institute of Electrical and Electronics Engineers Inc., Malaysia, pp. 254–258. https://doi.org/10.1109/SCOReD50371.2020.9250984, 9781728193175.

Izonin, I., Tkachenko, R., Kryvinska, N., Tkachenko, P., Greguš ml, M., 2019. Multiple linear regression based on coefficients identification using non-iterative SGTM neural-like structure. In: Lecture Notes in Computer Science (Including Subseries Lecture Notes in Artificial Intelligence and Lecture Notes in Bioinformatics), vol 11506. Springer Verlag, Ukraine, pp. 467–479. https://doi.org/10.1007/978-3-030-20521-8_39, 16113349.

Izonin, I., Tkachenko, R., 2022. An approach towards the response surface linearization via ANN-based cascade scheme for regression modeling in Healthcare. Proc. Comput. Sci. 198, 724–729. https://doi.org/10.1016/j.procs.2021.12.313, 18770509.

Jangra, M., Dhull, S.K., Singh, K.K., Singh, A., 2023. Denoising techniques for ECG arrhythmia classification systems: an experimental approach. In: Lecture Notes on Data Engineering and Communications Technologies, vol 175. Springer Science and Business Media Deutschland GmbH, India, pp. 1–13. https://doi.org/10.1007/978-981-99-3432-4_1, 23674520.

Krak, I., Kuznetsov, V., Kondratiuk, S., Azarova, L., Barmak, O., Padiuk, P., 2023. Analysis of deep learning methods in adaptation to the small data problem solving. In: Lecture Notes on Data Engineering and Communications Technologies, vol 149. Springer Science and Business Media Deutschland GmbH, Ukraine, pp. 333–352. https://doi.org/10.1007/978-3-031-16203-9_20, 23674520.

Krak, I., Barmak, O., Manziuk, E., 2022. Using visual analytics to develop human and machine-centric models: a review of approaches and proposed information technology. Comput. Intell. 38 (3), 921–946. https://doi.org/10.1111/coin.12289, 0824–7935.

Manziuk, E., Barmak, O., Krak, I., Mazurets, O., Skrypnyk, T., 2021. Formal model of trustworthy artificial intelligence based on standardization. In: CEUR Workshop Proceedings, pp. 190–197, 2853, 16130073.

Mochurad, L., Dereviannyi, A., Antoniv, U., 2021. Classification of X-ray images of the chest using convolutional neural networks. In: CEUR Workshop Proceedings, pp. 269–282, 3038, 16130073.

Mochurad, L., Shchur, G., 2021. Parallelization of cryptographic algorithm based on different parallel computing technologies. In: CEUR Workshop Proceedings, pp. 20–29, 2824, 16130073.

Nguyen Gia, T., Dhaou, I.B., Ali, M., Rahmani, A.M., Westerlund, T., Liljeberg, P., Tenhunen, H., 2019. Energy efficient fog-assisted IoT system for monitoring diabetic patients with cardiovascular disease. Future Generat. Comput. Syst. 93, 198–211. https://doi.org/10.1016/j.future.2018.10.029, 0167739X.

Pavliuk, O., Kolesnyk, H., 2023. Machine-learning method for analyzing and predicting the number of hospitalizations of children during the fourth wave of the COVID-19 pandemic in the Lviv region. J. Rel. Intell. Environ. 9 (1), 17–26. https://doi.org/10.1007/s40860-022-00188-z, 2199–4668.

Radiuk, P., Kovalchuk, O., Slobodzian, V., Manziuk, E., Barmak, O., Krak, I., 2022. Human-in-the-Loop approach based on MRI and ECG for healthcare diagnosis. In: CEUR Workshop Proceedings, pp. 9–20, 3302, 16130073.

Saravanan, V., Aneja, I., Yang, H., Pillai, A.S., Singh, A., 2021. Impact of big data in healthcare system-a quick look into electronic health record systems. In: Machine

Learning and the Internet of Medical Things in Healthcare. Elsevier, United States, pp. 251–262. https://doi.org/10.1016/B978-0-12-821229-5.00009-4, 9780128212295.

Semkovych, V., Shymanskyi, V., 2023. Combining OCR methods to improve handwritten text recognition with low system technical requirements. In: Lecture Notes on Data Engineering and Communications Technologies, vol 158. Springer Science and Business Media Deutschland GmbH, Ukraine, pp. 693–702. https://doi.org/10.1007/978-3-031-24475-9_56, 23674520.

Shafay, M., Ahmad, R.W., Salah, K., Yaqoob, I., Jayaraman, R., Omar, M., 2023. Blockchain for deep learning: review and open challenges. Cluster Comput. 26 (1), 197–221. https://doi.org/10.1007/s10586-022-03582-7, 15737543.

Shakhovska, N., Yakovyna, V., Chopyak, V., 2022. A new hybrid ensemble machine-learning model for severity risk assessment and post-COVID prediction system. Math. Biosci. Eng. 19 (6), 6102–6123. https://doi.org/10.3934/mbe.2022285, 1551–0018.

Soltanzadeh, P., Hashemzadeh, M., 2021. RCSMOTE: range-controlled synthetic minority over-sampling technique for handling the class imbalance problem. Inf. Sci. 542, 92–111. https://doi.org/10.1016/j.ins.2020.07.014, 00200255.

Tkachenko, R., 2022. An integral software solution of the SGTM neural-like structures implementation for solving different data mining tasks. In: Lecture Notes on Data Engineering and Communications Technologies, vol 77. Springer Science and Business Media Deutschland GmbH, Ukraine, pp. 696–713. https://doi.org/10.1007/978-3-030-82014-5_48, 23674520.

Tkachenko, R., Tkachenko, P., Izonin, I., Tsymbal, Y., 2018. Learning-based image scaling using neural-like structure of geometric transformation paradigm. In: Studies in Computational Intelligence, vol 730. Springer Verlag, Ukraine, pp. 537–565. https://doi.org/10.1007/978-3-319-63754-9_25, 1860949X.

Tkachenko, R., Izonin, I., 2019. Model and principles for the implementation of neural-like structures based on geometric data transformations. In: Advances in Intelligent Systems and Computing, vol 754. Springer Verlag, Ukraine, pp. 578–587. https://doi.org/10.1007/978-3-319-91008-6_58, 21945357.

Tkachenko, R., Izonin, I., Vitynskyi, P., Lotoshynska, N., Pavlyuk, O., 2018. Development of the non-iterative supervised learning predictor based on the Ito decomposition and SGTM neural-like structure for managing medical insurance costs. Data 3 (4). https://doi.org/10.3390/data3040046, 2306–5729.

Tolstyak, Y., Havryliuk, M., 2022. An assessment of the transplant's survival level for recipients after kidney transplantations using cox proportional-hazards model. In: CEUR Workshop Proceedings, pp. 260–265, 3302, 16130073.

Tolstyak, Y., Chopyak, V., Havryliuk, M., 2023. An investigation of the primary immunosuppressive therapy's influence on kidney transplant survival at one month after transplantation. Transpl. Immunol. 78. https://doi.org/10.1016/j.trim.2023.101832, 09663274.

Uci, 2023. Machine Learning Repository N.

Vu, D.-H., 2022. Privacy-preserving Naive Bayes classification in semi-fully distributed data model. Comput. Secur. 115. https://doi.org/10.1016/j.cose.2022.102630, 01674048.

CHAPTER 10

Legal and regulatory issues related to AI in healthcare

Anshita Gera[1], Anuradha Dhull[1], Akansha Singh[2] and Krishna Kant Singh[3]
[1]Department of CSE, The NorthCap University, Gurugram, Haryana, India; [2]School of CSET, Bennett University, Greater Noida, Uttar Pradesh, India; [3]Delhi Technical Campus, Greater Noida, Uttar Pradesh, India

Introduction

A broad area of computer science called artificial intelligence aims to create intelligent computers or devices that can carry out activities that would typically need human intellect. These activities include a broad range of skills, such as learning, voice recognition, problem-solving, and decision-making. AI systems are made to analyze large amounts of data, spot trends, and use machine learning techniques to keep becoming better at what they do. The two primary categories of AI are narrow or weak AI, designed for specific tasks, and general or strong AI, aiming to exhibit human-like intelligence across diverse domains. The integration of AI technologies into various sectors, such as healthcare, finance, and autonomous systems, has inspired exceptional advancements, but it also raises ethical, legal, and societal considerations that necessitate thoughtful exploration and regulation(Prakash et al., 2022).

The rise of AI in healthcare

Innovative uses of AI in healthcare, such as natural language processing, predictive analytics, and diagnostic imaging, are becoming increasingly common and the medical fields that are most benefited from AI can be referred and better understood from Fig. 10.1 mentioned below. These developments include better data extraction from healthcare records, more effective diagnostic tools, and customized treatment regimens (Bohr and Memarzadeh, 2020). Nevertheless, its incorporation presents fresh

Responsible and Explainable Artificial Intelligence in Healthcare
ISBN 978-0-443-24788-0
https://doi.org/10.1016/B978-0-443-24788-0.00010-8

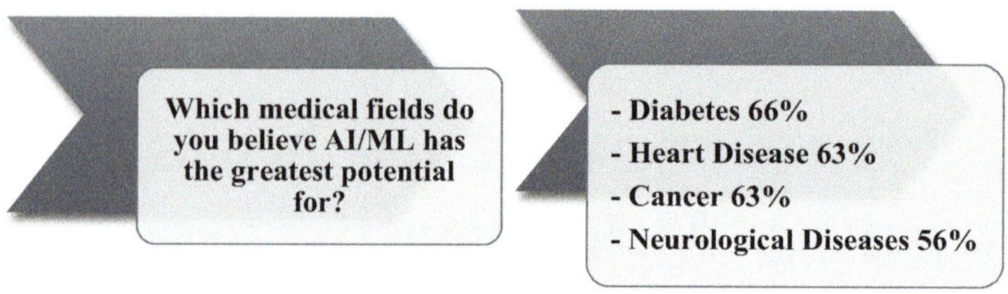

Figure 10.1 Chronic health conditions expected to benefit most from AI/ML.

difficulties for current legal structures. Concerns of responsibility, transparency, and culpability are brought up by the opacity of some AI algorithms, particularly when unfavorable things happen. It takes a so patient health informationsticated grasp of the changing legal landscape in healthcare AI to navigate these obstacles while promoting innovation and guaranteeing patient safety (Krittanawong, 2018).

AI in healthcare market trends

The market trends of AI in healthcare have witnessed a remarkable ascent in recent years. We can infer from the statistics given in Fig. 10.2 below, that the use of AI in healthcare has grown significantly; by 2021, the worldwide industry is expected to reach $11 billion. By 2030, it is predicted to have increased to around $188 billion, with a compound annual growth rate of 37% between 2022 and 2030. This increase is a result of the broad

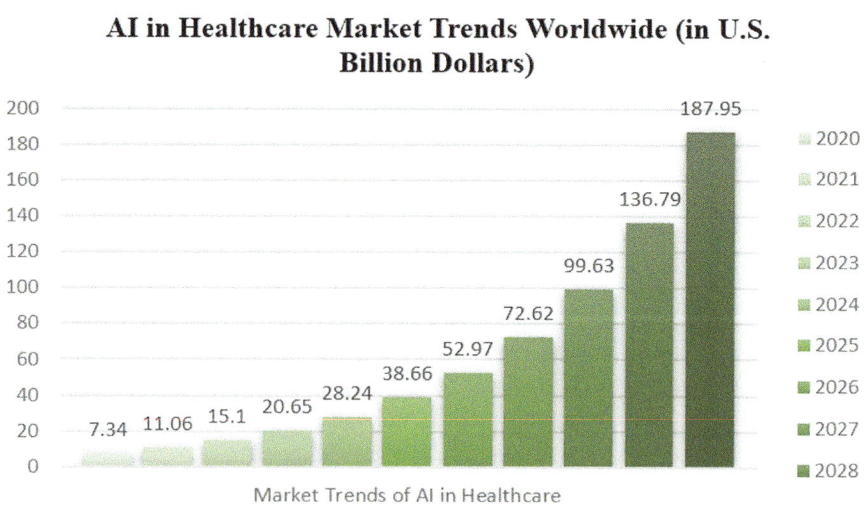

Figure 10.2 Market trends of AI in healthcare.

use of AI technology and a paradigm shift toward creative solutions that use artificial intelligence to improve global healthcare outcomes (Vyas, 2023).

Legal and regulatory landscape

The legal and regulatory environment that surrounds AI in healthcare is complex and ever-changing. The healthcare industry's current structures and laws offer a starting point, but there are obstacles in their capacity to be modified to incorporate AI applications. The United States' Health Insurance Portability and Accountability Act (HIPAA) and the European Union's General Data Protection Regulation (GDPR) (Feld, 2005) are an example of a current policy that focuses largely on data security and privacy, although it may not have rules that specifically address the complexities of AI in healthcare. Finding these gaps is essential to comprehending the constraints of the laws that are in place and developing focused strategies to guarantee data integrity, patient safety, and ethical AI activities. Closing these regulatory loopholes will be necessary as AI continues to progress in the healthcare industry in order to promote innovation and uphold the highest standards of morality and legality (Khan, 2024).

Legal and ethical problems raised

Artificial Intelligence on one hand contributed a lot in different domains where as on other hand raised a handful of legal and regulatory issues in a variety of fields. To discuss this in brief refer to Fig. 10.3 below where key problems raised in different AI domains are put.

Importance of resolving legal and regulatory matters

Addressing legal and regulatory issues in healthcare AI is crucial due to potential risks such as compromising patient safety and data privacy (Naik et al., 2022). There is a greater chance of jeopardizing patient safety, data privacy, and the general integrity of healthcare systems in the absence of strong frameworks designed expressly to handle the complexities of AI in healthcare. Legal ambiguities might make it difficult to assign blame when AI-related mistakes occur. Furthermore, in the lack of clear regulations, stakeholders may be discouraged from pursuing novel AI technology in healthcare due to concerns about legal ramifications and compliance. In order to create an atmosphere where AI may smoothly incorporate into healthcare procedures, protecting patient welfare and upholding the confidence of the public and healthcare practitioners alike, proactive resolution of these concerns is imperative (Harvey and Gowda, 2021).

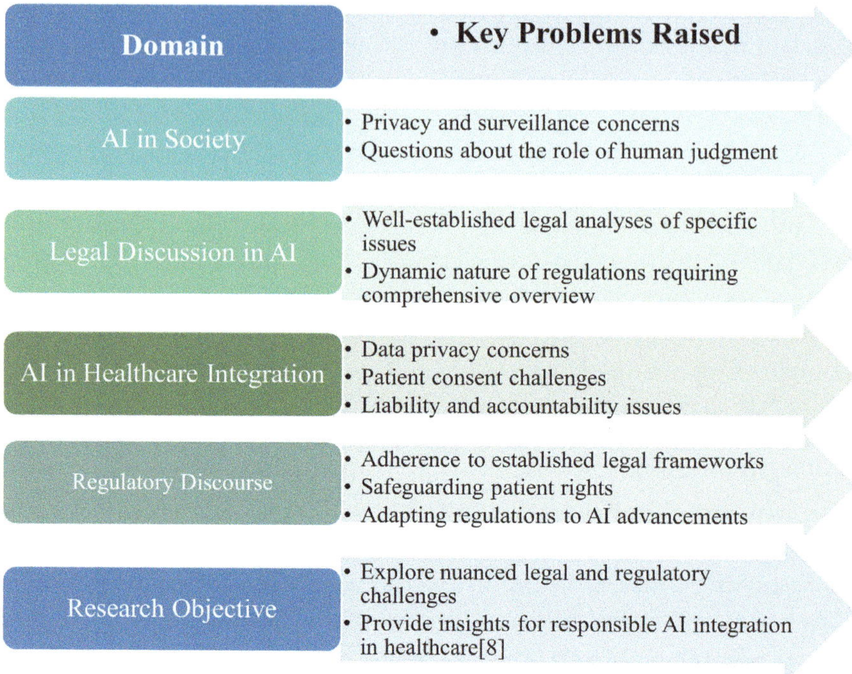

Figure 10.3 Key problems arisen in different AI domains (Schönberger, 2019).

Research Questions:

Q-1: What is the contribution of AI in the field of Healthcare, its Benefits, Challenges and Market Trends?

Q-2: What are the Legal and Regulatory Issues faced with AI in Healthcare?

Q-3: What are the mainstream use cases of AI in Healthcare?

Benefits of AI in healthcare

As AI-based technologies are applied more widely in the healthcare sector, a vast range of new opportunities arise. Here, a few of the most significant ones are covered (Chaieb et al., 2023). The key benefits of AI in healthcare are summarized in table Table 10.1.

Challenges faced with legal issues of AI in healthcare

AI provided its utilization and applications in various fields but it also gave some challenges that are required to be dealt with. Here, some of most prominent ones are addressed (Pesapane et al., 2018). The challenges faced due to legal issues of AI in Healthcare are summarized in Table 10.2.

Table 10.1 Key benefits of AI in healthcare.

S. no.	Key benefits	Description
1.	Enhanced disease management	AI, such as IBM Watson, improves patient care by aiding in disease management, particularly in remote areas, and has the potential to match or surpass medical professionals in precision.
2.	Increased patient involvement and engagement	Health coaching apps such as noom empower individuals to adopt healthier habits and lower chronic disease risks, requiring commitment for achieving set health goals and improving overall patient involvement.
3.	High quality of service and reduced errors	AI systems, integrating innovative algorithms and deep learning, enhance diagnostic accuracy, promising improved quality in healthcare services and reduced error rates.
4.	Additional functionality and lower medical expenses	AI technologies, including pill-cams and diagnostic algorithms, expedite processes such as leukemia testing, enhancing productivity and reducing costs. AI optimization of workflows and integration of AI-powered chatbots contribute to cost-effective healthcare delivery.

AI in healthcare: A gift or a threat?

Cath (2018) presented in her research article that artificial intelligence is starting to appear in more and more areas of our life, from the essential—such as banking, law enforcement, healthcare, and humanitarian aid—to the trivial—such as dating. The accompanying Fig. 10.4 also provides insight into the implementation of AI in healthcare. The use of artificial intelligence has the potential to improve social and economic

Table 10.2 Challenges faced due to legal issues of AI in healthcare.

S. no.	Challenges	Description
1.	Intellective rights	- LLMs in medical studies pose challenges for IP rights.- risks include potential loss of rights.- healthcare professionals need robust rules, processes, and training for IP development in the healthcare sector.
2.	Ignorance and responsibility due diligence	- Legal implications for AI-related injuries involve various stakeholders.- difficulty in proving causation within dynamic machine learning systems (Gerke et al., 2020).- duty of care concerns for healthcare professionals using AI, highlighting the importance of patient disclosure and education.
3.	New issues from therapeutic goods act changes	- TGA cybersecurity guidelines address risks in software based medical devices.- digitalization introduces new risks such as privacy breaches and modified device features causing harm.- reforms emphasize the need for precise standards and new categorization regulations for software-based medical devices (Opderbeck, 2019).

well-being as well as the exercise of legal and human rights. This covers machine learning methods and embodied AI in robots. The need for transparent, egalitarian, and responsible AI development and regulation is growing. What frameworks and how may this be achieved? This is one of the primary subjects addressed in this special edition. The legitimate, scientific, and legal-regulatory challenges of developing governance frameworks for AI systems are examined in-depth by eight authors. It also gives a brief overview of recent developments in AI governance and indicates how much of the agenda for defining AI regulations, ethical frameworks, and technological approaches has already been defined, in addition to making some specific recommendations to advance the conversation on AI governance.

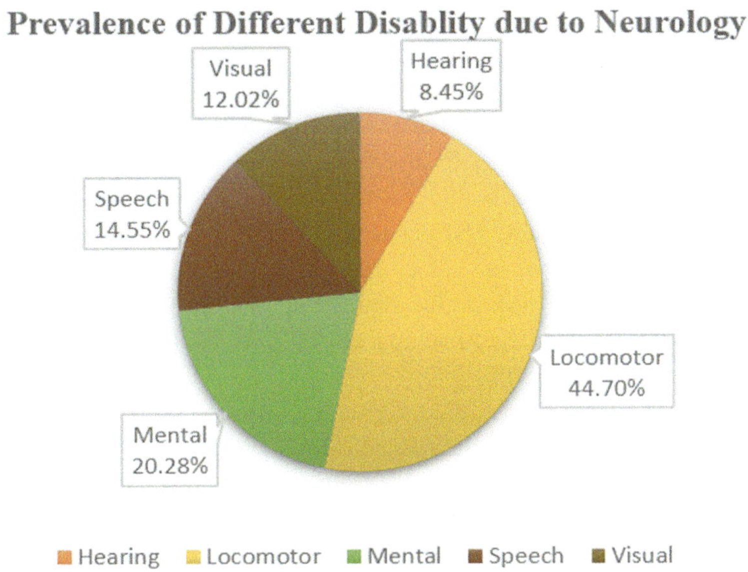

Figure 10.4 Prevalence of different diseases due to neurology.

WIS AI (2018) also provided insights on the same in his paper that the planning and distribution of resources for health and social care services may benefit from the application of AI. To improve cost-effectiveness, Harrow Council is experimenting with the IBM Watson Care Manager system. It matches individuals with a caregiver who meets both their needs and their designated care budget. It also claims to offer advice on how to make better use of care management resources and develops customized care plans (WIS AI, 2018).

Gerke et al. (2020) presented in her research that many think artificial intelligence will completely change healthcare, from workflow optimization in hospitals to clinical applications in fields such as imaging and diagnostics to the assessment of a patient's symptoms using health applications. The AI health sector is anticipated to grow quickly in the years to come; according to one study, the market is likely to grow by more than 10 times between 2014 and 2021. This increase is accompanied by a number of problems, thus it's critical that AI be applied in the healthcare system in a way that abides by the law and moral principles. It outlined the moral and legal issues that artificial intelligence in healthcare raises and provide solutions (Gerke et al., 2020).

Molnár-Gábor (2019) discussed about the benefits and drawbacks of integrating AI into healthcare are highlighted, especially when it comes to defining ideas of health and illness. An application of ethical principles and liability rules is necessary to address potential errors in judgment and responsibility in the use of medical AI. A review of existing AI use cases indicates that roles are not well defined, leaving those who are impacted to manage risks on their own. The position of the physician is still unclear, but by viewing the physician-patient connection as a wide contractual duty, it may be possible to strengthen the established standing of physicians in this changing environment by improving ethical and legal norms in AI applications and clarifying duties.

Naik et al. (2022) researched about the number of legal, ethical and regulatory challenges that society faced due to artificial intelligence, such as issues of prejudice or intolerance, confidentiality and monitoring, the theoretical issue of what function human judgment serves, and other aspects of the medical profession. The introduction of modern digital technologies has raised concerns about their being a new source of inaccurate information and data breaches. When errors in process or protocol occur in the healthcare industry, the patient who is the victim of the error may suffer grave repercussions. Remembering this is important since patients often interact with physicians during times in their lives when they are most vulnerable. There is currently an absence of specific laws in health care institutions to address potential ethical and legal issues arising from the use of AI. In an effort to address these important concerns, this analysis emphasizes the necessity of algorithmic openness, privacy, and protection of all parties involved, as well as cybersecurity of related vulnerabilities (Naik et al., 2022).

Mainstream use cases involving AI's legal and regulatory concerns in healthcare

A number of complex legal challenges are raised by the integration of AI across several medical domains, such as reproductive health, juvenile justice, college mental health, adolescent healthcare, and patient data protection etc.

Furthermore, the use of AI in wearables, preventative healthcare, and the fight against fraud and exploitation has created new legal difficulties that call for extensive regulatory frameworks to ensure the ethical and responsible application of AI in a range of healthcare settings. All the above mentioned use cases are discussed in detail below.

Legal and ethical concerns in the realm of reproductive health within the field of neurology

Kirkpatrick et al., (2022) presented in the research article byinterviewing pediatric neurologists regarding sexual and reproductive health for WWE with intellectual disability. Inthe paper they also researched on the Prevalence of Different kinds of Disability that arises due to Neurology which can be seen in Fig. 10.4. Research on the difficulties paediatric neurologists encounter when treating young adults and adolescents with epilepsy (WWE), especially those who have intellectual disabilities, regarding their sexual and reproductive health reveals a complicated landscape that is relevant to legal and regulatory issues in the healthcare sector.

The divergent views of pediatric neurologists about the ability of people with epilepsy to engage in sexual activity and parent reflect the moral conundrums AI encounters when evaluating complex medical requirements. The debates around AI's ability to navigate legal frameworks for permission in healthcare decisions are similar to those surrounding the level of discussion for capacity for sexual consent.

The recommendations made by the study for clinical standards and more training for pediatric neurologists are consistent with the need for AI systems to comply with strong regulatory frameworks. Similar to the study's recommendations, standardizing procedures and guaranteeing ongoing training for AI algorithms are crucial in the field of AI in healthcare in order to meet the special requirements of people with intellectual impairments.

The suggestion for multidisciplinary cooperation in AI applications amongst neurologists, primary care physicians, and women's health professionals is in line with this idea. In the field of artificial intelligence, cooperation amongst different stakeholders—such as developers, medical professionals, and regulatory agencies—becomes essential to guarantee thorough and knowledgeable decision-making.

Arguments over the involvement of pediatric neurologists in the decision-making process around contraception bear similarities to the moral dilemmas AI faces when handling delicate medical decisions. Questions of appropriateness and ethical concerns are raised by the different recommendations and ethical considerations across physicians, which are similar to the difficulties AI may have while providing individualized healthcare recommendations.

The study supports the need for ethical AI standards by recommending clinical practice recommendations that are specific to the illness, developmental stage, and cognitive state of people with intellectual impairments. The study's emphasis on customized clinical recommendations is reflected in the development of AI systems in the healthcare sector that are sensitive to a variety of patient profiles, maintain anonymity, and actively coordinate amongst various healthcare system components. The challenges outlined in the study shed light on legal, regulatory, and ethical considerations in providing sexual and reproductive health care for individuals with epilepsy and intellectual disabilities. Aligning these findings with the context of AI in healthcare highlights the need for robust legal frameworks, ethical guidelines, and interdisciplinary collaboration to ensure responsible and personalized healthcare solutions for vulnerable populations (Raine, 2023).

HIV-related public health, legal, and moral concerns pertaining to expectant mothers and new-borns

Tessmer-Tuck et al. addressed a compelling narrative within the field of AI in healthcare, where advanced technologies intersect with legal, ethical, and public health challenges. The integration of artificial intelligence into healthcare systems is intended to enhance efficiency, improve patient outcomes, and streamline processes (Tessmer-Tuck et al., 2014). The urgency of early intervention to prevent pediatric AIDS cases aligns with the goals of AI applications in healthcare, aiming to enhance early detection and diagnosis for improved patient outcomes. The legal complexities surrounding pregnant women who decline AI-driven medical procedures parallel the evolving discussions around patient autonomy and consent in the context of AI applications. The study underscores a woman's right to refuse medical testing and treatment, emphasizing the need for legal frameworks that respect individual choices, even in the face of potential benefits to the patient.

The absence of a definitive United States Supreme Court ruling in the case mirrors the current state of legal frameworks for AI in healthcare. The rapidly evolving nature of AI technologies presents challenges in adapting existing regulations to keep pace with advancements. The call for clinicians to stay vigilant about state-specific developments resonates with the dynamic regulatory landscape surrounding AI in healthcare, where ongoing legal challenges and changes are the norm.

The focus on an HIV-positive pregnant woman declining AI-driven testing and treatment adds complexity to discussions around patient refusal of AI interventions. Legal intervention becomes necessary, highlighting the rarity of such cases but also underscoring the potential for legal challenges as AI becomes more integrated into healthcare systems. This emphasizes the importance of establishing clear guidelines and regulations to navigate situations where patients refuse AI-driven medical interventions.

The tension between a mother's right to refuse medical procedures and the desire to provide optimal care for the fetus in the case study mirrors the ethical considerations surrounding AI applications. Striking the right balance between respecting individual autonomy and leveraging AI for collective public health benefits is a central challenge. Legal and regulatory frameworks must evolve to accommodate these nuanced ethical considerations in the deployment of AI technologies. This case provides a valuable lens through which to explore the legal and regulatory dimensions of AI in healthcare. It underscores the need for robust legal frameworks, ethical guidelines, and ongoing collaboration between healthcare practitioners, technologists, and regulatory bodies. Addressing these issues is crucial to ensuring the responsible and effective integration of AI in healthcare, respecting patient autonomy while harnessing the potential benefits of advanced technologies (Faden and Geller, 1991).

Juvenile justice: Legal concerns, rights, and ethics for mental health

Thomas et al. provided a valuable insights in the field of AI in context of mental health and the juvenile justice system in which he presents a complex web of legal, ethical, and individual rights problems that are similar to those in the healthcare industry. In order to protect the rights of juvenile offenders and uphold ethical integrity, it is critical to synchronize AI breakthroughs with legal and regulatory requirements as they become more widely used in mental health therapies.

The juvenile justice system's legislative framework places significant emphasis on the constitutional rights of minors. The rights of children to tailored and restorative care are recognized by the Supreme Court, and this includes AI applications in mental health. AI-driven mental health treatments are required by law to be in line with the developmental requirements of minors and to prioritize rehabilitation above punitive measures. As artificial intelligence (AI) is used more and more to help mental health treatment in the context of juvenile justice, it is imperative that these constitutional norms are followed.

Privacy rights and confidentiality are also major issues in the legal system. To preserve the privacy of juvenile mental health records, AI systems handling sensitive mental health data must abide by current regulatory frameworks, such as the Health Insurance Portability and Accountability Act (HIPAA) in the United States. The ethical use of AI in mental health treatment requires finding the ideal balance between maintaining anonymity and providing the legal system with pertinent information (Thomas, 2020).

Ethical considerations in the use of AI technologies within juvenile justice mental health care parallel those in traditional healthcare contexts. The cultural competence of AI algorithms becomes paramount, ensuring that they are sensitive to the diverse backgrounds and experiences of juvenile offenders. Ethical guidelines underscore the responsibility of developers and practitioners to address potential biases within AI algorithms that could perpetuate disparities in mental health care outcomes.

Informed consent, a fundamental ethical principle, takes on new dimensions in the context of AI. While juveniles may lack the legal authority for independent consent, involving them in decisions regarding AI-driven interventions to the extent possible upholds their autonomy. Transparency about the functioning of AI systems, the potential outcomes, and the limitations of these technologies becomes crucial to ethical practice.

Furthermore, in the context of AI-driven mental health, it is morally necessary to promote the interests of young offenders. In order to make sure that new technologies support good results, are in line with rehabilitative aims, and give adolescents in the juvenile justice system priority, mental health practitioners and AI developers must work together. To summarise it was also examined that the integration of AI in mental health into the juvenile justice system necessitates a careful examination of ethical issues, individual rights protection, and legal compliance. The growing field of juvenile justice mental health treatment necessitates continued collaboration across legal, regulatory, and ethical frameworks in order to strike a balance between utilizing AI's advantages and protecting the rights and welfare of young offenders (Koocher and Kinscherff, 2016).

Concerns about law and ethics in college mental health

Glied et al. studied the issues that were raised in students and affected their mental health and provided study on the implementation of AI in college mental health services presents a complicated environment that requires cautious navigation due to the convergence of legal and ethical constraints. It has been seen that while AI has the potential to improve accessibility and efficiency in the healthcare industry, especially in the area of mental health services on college campuses, it also poses a number of ethical and legal issues.

Privacy and the safeguarding of private health information are two major legal concerns. In the United States, the Health Insurance Portability and Accountability Act (HIPAA) as discussed previously in the former use case of "Juvenile Justice", governs the privacy of mental health information pertaining to college students. Therefore, in order to protect students' privacy, any AI system used in mental health services needs to abide by strict privacy rules. To avoid unwanted access and possible breaches, institutions need to put strong data security measures in place and make sure AI applications abide by HIPAA rules and other pertinent data protection legislation.

The ethical issues surrounding the application of AI to college mental health services are also quite important. Fairness, openness, and impartial decision-making must be given top priority in the creation and application of AI systems. If the AI system demonstrates prejudices based on racial, gender, or socioeconomic background, it might perpetuate gaps in mental health care, raising ethical questions. In order to guarantee that AI technologies do not unintentionally contribute to prejudice or inequality, addressing these biases needs constant attention, clear algorithmic design, and frequent audits.

Another ethical factor considered here is informed consent. When using AI-powered mental health services, college students should be properly educated about the nature of the technology, its limits, and any possible effects on their care. Upholding the values of autonomy and openness means making sure that students are fully aware of how AI algorithms work, how their data will be used, and the possible results of AI-assisted interventions.

Moreover, human supervision plays a critical role in AI-driven mental health therapies. The ethical framework should prioritize a collaborative approach wherein human mental health experts stay actively involved in the decision-making process, even when AI may offer vital insights and support. This guarantees that technology efficiency won't come at the expense of the human touch that is essential to mental health treatment.

As AI becomes more widely used in college mental health services, organizations need to create strong rules and processes that adhere to ethical and legal guidelines. This includes putting in place thorough training programs for employees, dealing with potential biases, and routinely assessing how AI affects patient outcomes. The continued difficulty of reconciling the potential of AI-enhanced mental health treatments with legal and ethical issues calls for a dedication to fairness, transparency, and college students' well-being (Glied et al., 2019).

Treating adolescent patients in AI in healthcare: Legal and ethical considerations

Sisk et al. conducted a research considering Adolescent Patients and observed that the medical professionals face special legal and regulatory requirements while caring for teenage patients, and the use of AI makes matters more difficult. Consent, confidentiality, and the developing ability of adolescents to make medical decisions are all factors to be taken into account while providing healthcare to adolescents. Artificial intelligence technologies add layers of complexity, therefore legal and regulatory frameworks need to be carefully examined (Sisk et al., 2019).

Informed consent is a major legal consideration when treating adolescents. Teenagers under a specific age may need parental or guardian approval before receiving medical treatment in various jurisdictions. But when they become older, they could be given the freedom to choose their own healthcare providers. The difficult part is figuring out whether a teenager is old enough to give informed consent—a decision that varies depending on the state. This problem is made more complicated by AI applications in healthcare, particularly those that include making judgments about diagnosis or therapy. Getting legitimate permission from teenagers or their guardians means making sure they comprehend the nature of AI-driven interventions.

Another essential component is confidentiality, with regulatory structures intended to safeguard teenage patients' privacy. When it comes to delicate subjects such as mental or reproductive health, health information privacy rules frequently control when and how healthcare practitioners can communicate medical information. There are worries over data security and the possibility of unintentional disclosures when using AI algorithms to handle and analyze patient data. Adolescent healthcare uses AI technology, therefore adhering to existing privacy rules becomes crucial.

An important factor in determining how the adolescent healthcare environment is shaped is regulatory monitoring. Standards and guidelines are established by regulatory agencies to guarantee the effectiveness and safety of medical treatments. It is crucial to follow current laws and create certain guidelines for AI applications when incorporating AI into adolescent healthcare. Regulatory clearance and patient confidence are contingent upon AI technology adhering to safety and ethical criteria.

Moreover, because AI technologies are developing quickly, it is necessary to change regulations often to stay up to current as presented by Sharko et al. in his research. Healthcare professionals and AI developers must remain up to date on regulatory developments in this ever-changing landscape in order to maintain legal compliance.

Careful thought and adherence to changing legal frameworks are necessary to strike a balance between the autonomy and rights of teenage patients and the difficulties posed by AI applications. Responsible and ethical integration of AI in adolescent healthcare requires defined criteria for informed consent, confidentiality, and regulatory landscape navigation. In order to handle these complex concerns and promote the moral use of AI technologies in the treatment of teenage patients, it is imperative that regulators, legal professionals, and healthcare practitioners work together continuously (Sharko et al., 2022).

The social, legal, and ethical implications of AI to the treatment of breast cancer

Carter et al. under careful observation addressed that as a reflection of the dynamic nature of AI applications in healthcare, the integration of AI in breast cancer diagnosis raises a number of legal and regulatory problems (Carter et al., 2020). The field of breast cancer screening is changing, and although AI-based technologies offer increased efficacy and precision, there are important ethical and regulatory ramifications.

The accuracy and dependability of AI models used in breast cancer screening is one of the main legal concerns. The wide range of accuracy recorded by several research (69.2%—97.8%) begs the issue of how reliable these AI systems are. To guarantee consistent and dependable performance across a variety of datasets, legal frameworks must take into account the requirement for defined assessment criteria and validation procedures.

There are difficulties since many studies are retrospective in nature and because datasets may contain biases due to the abnormally high number of mammograms with cancer diagnoses. Legal issues should include policies that promote diversity, minimize biases, and enable public reporting of dataset features in order to improve the generalizability of AI models for the diagnosis of breast cancer.

Legal complications arise when AI-based breast cancer screening is commercialized outside of traditional healthcare facilities, especially by start-ups. These businesses could advertise directly to customers, focusing on test methods that are still in their infancy and attracting those in the age range below which mammography is advised. Legal frameworks must closely monitor marketing strategies to guarantee open communication, moral behavior, and consumer rights protection.

Collaborations between start-ups and larger technology providers highlight the need for clear legal agreements and responsibilities. Regulations should address issues related to technology transfer, intellectual property, and data ownership to prevent disputes and ensure a fair and ethical business environment.

Houssami et al. also provided insights in the matters of Patient's data privacy, that it is a paramount legal consideration in AI applications. As breast cancer detection involves sensitive health information, adherence to existing data protection laws, such as HIPAA (discussed earlier), is crucial. Legal frameworks must ensure that AI systems comply with stringent data security measures to prevent unauthorized access and protect patient confidentiality. Ultimately, addressing the unique issues offered by these technologies calls for a proactive strategy given the legal context around the use of AI in breast cancer screening. Regulatory agencies ought to work together with interested parties to provide uniform assessment procedures, encourage openness in reporting, protect patient confidentiality, and guarantee moral advertising. In the area of breast cancer screening, the legal framework must change as AI develops in order to maintain patient rights while promoting innovation (Houssami et al., 2019).

Safeguarding patients' personal information

Roy et al. discussed majorly on the issue of allowing AI access to private medical information presents serious obstacles on a national and international scale, highlighting the necessity of striking a careful balance between patient populations' biological variety, medical ethics, and the availability of representative datasets in his research and studied that this becomes especially important in the case of uncommon disorders, for which research and development require access to a complete medical database.

The Convention for the Protection of Individuals with respect to Automated Processing of Personal Data (1981, Strasbourg) now forms part of the international legal framework for the protection of personal data.

This convention defines methods for data protection and processing as well as basic ideas, guidelines, and requirements for the handling of personal data. In 2013, Russia accepted the convention in accordance with international norms, leading to the enactment of Federal Law No. 152-FZ of July 27, 2006, "On Personal Data," which established broad guidelines and regulations for the handling of personal data (Roy, 2022).

Murdoch et al. also addressed this issue and after careful observation presented that legal adjustments are required to establish advantageous conditions for the responsible access of AI and robotics developers to a variety of data kinds, including those gathered by governmental and medical institutions, as the work at hand entails improving data availability and quality. To control access to personal data for scientific research, AI education, and technological advancement, specific modalities or criteria should be put in place. These requirements must respect the right to privacy and incorporate depersonalization mechanisms while maintaining a balance between promoting research and protecting individual interests.

It is found crucial to be clear when asking for consent to process personal data for scientific study in AI-related disciplines. This holds true for fields such as sociology, ecology, and health, among others. The legal requirements for utilizing AI and robots to organize identity should also be specified, with a focus on protecting personal information and upholding people's right to privacy.

Any extra requirements should guarantee enhanced protection of persons' rights since personal data is sensitive. It is advisable to contemplate modifying the current personal data laws in different states to address the distinct obstacles presented by artificial intelligence in the healthcare sector. These changes have to be in line with how healthcare AI is developing and guarantee a peaceful coexistence of technology innovation with the defense of peoples' rights and privacy. By doing this, the legal and regulatory framework may promote responsible innovation and protect patient interests throughout the world by more effectively addressing the issues resulting from the junction of AI and healthcare (Murdoch, 2021).

"Doctor ChatGPT": LLMs and a request for regulation

Barrington et al. studied Large Language Models and the issues accompanied with it and after cautious observation it was found that the emergence of Generative AI chat tools, such as as "Tessa," the AI-powered LLM chatbot from the National Eating Disorder Association (NEDA)

(Barrington et al., 2023), offers significant problems as well as potential prospects for the healthcare industry, if controlled properly, these technologies may be very helpful to doctors and patients in addressing different medical issues, creating treatment plans, and locating resources. But given the possible hazards and uncertainties, the state of affairs today begs the important question: is it wise to use these instruments for medical purposes?

NEDA created Tessa with the intention of fostering wellbeing and offering support to anyone impacted by eating disorders. Although Tessa's goals were admirable, she displayed what is known as "hallucinations" in tech speak — instances of incorrect replies. This is something that many AI chatbots have to deal with. This problem is not exclusive to NEDA; other conversation platforms driven by generative AI have also struggled to provide accurate and trustworthy information. Giving people inaccurate or misleading information can have serious repercussions, putting them at risk, especially when it comes to medical advice.

A pivotal moment has arrived in the legal and regulatory environment around the use of AI in healthcare, particularly with regard to chatbots that provide medical advice, regulators have taken notice of the possible ramifications that might result from errors in chatbot-generated medical advice, especially when these tools are supported by groups in the healthcare sector such as NEDA. Because AI systems have the potential to negatively effect patient outcomes, accuracy and dependability of information given in a healthcare context are critical are some of the important parameters that were discussed in the study presented by Wang et al.

Wang et al. also addressed that regulators are realizing more and more how important it is to provide precise rules and specifications for the application of AI in healthcare. In order to guarantee the appropriate use of these tools, it is becoming increasingly important to solve the legal and regulatory gaps as AI technologies advance. There are concerns about responsibility, liability, and the possible fallout from medical advice given by AI-powered chatbots that isn't entirely correct.

Protecting patient interests and promoting innovation must be balanced delicately in the legal framework. This entails outlining the obligations of healthcare institutions that make use of these AI technologies, setting criteria for precision and dependability, and specifying the possible legal ramifications in the event that inaccurate information is provided.

In conclusion, careful consideration of legal and regulatory frameworks is required when using Generative AI conversation tools in healthcare, as demonstrated by Tessa's situation. In order to ensure that these technologies can achieve their promise in improving healthcare practices while upholding the highest standards of accuracy and dependability, it is imperative that proactive actions be taken to limit risks in light of the developing landscape. In the field of AI-driven healthcare solutions, regulators are positioned to be major players in defining this landscape by highlighting the requirement of a harmonic junction between patient safety and technology innovation (Wang et al., 2023).

Legal and regulatory issues with IOT technology and wearables

Ramasamy et al. researched IOT Technology and provided a major contribution in this sector and addressed the legal and regulatory issues that are raised with it, in which he observed that modern healthcare delivery greatly depends on wearable gadgets, which have developed from basic pedometers to patient health information equipped systems that may identify chronic diseases and emergency crises. These gadgets produce a plethora of data that has never been seen before, providing faster and more accurate patient care choices. Furthermore, wearable technology encourages better involvement in the healthcare process by facilitating more contacts between patients and healthcare professionals. Still, patient data-driven algorithms—which include demographic patient health information and private medical data—are the source of these technologies.

National standards were developed by federal law (Ramasamy et al., 2022), to protect patient health information against unauthorized disclosure or use. The main line of defence against risks relating to wearable device collection and transfer of sensitive patient health information is these restrictions. Healthcare organizations must adhere to the regulations set out by HIPAA and state legislation in order to secure patient data. In a September 2022 presentation, the U.S. Department of Health and Human Services (HHS) Office of Information Security stressed the importance of taking important precautions in response to the rising issues around wearable devices. In order to stop patient health data from wearable devices from being intercepted, this involves promoting strong security methods including end-to-end encryption, entire disk encryption, and blanket multifactor authentication.

A recent class action lawsuit against OpenAI, the company that created ChatGPT, in the Northern District of California serves as an example of the legal issues that have arisen in the field of AI data gathering and usage. Based on data scraping of social media comments, as researched by Awotunde et al. in his research on IOT technology that chat logs, cookies, contact details, login passwords, and financial information, the case claims infringement of privacy rights. This case establishes a precedent for potential consequences in the context of abuse and highlights the growing legal attention over AI-related data activities.

Awotunde et al. also looked into the matter of present intricate legal and regulatory challenges that are arisen by the nexus of healthcare, AI, and wearable technology and maintaining the security and integrity of patient data requires strict adherence to regulatory rules and regulations. The increasing integration of wearable gadgets into healthcare procedures necessitates the evolution of regulatory frameworks to tackle new issues. Maintaining patient privacy while utilizing wearable technology is a delicate balance that requires ongoing work on the part of regulatory bodies and healthcare institutions. Even if wearable technology has the potential to revolutionize healthcare, integrating it raises legal and regulatory issues. To preserve data security standards and safeguard patient privacy, the changing environment necessitates more awareness. Legislators, healthcare professionals, and technology developers must work together to negotiate the complex legal and ethical implications of artificial intelligence in healthcare as technology develops (Awotunde et al., 2021).

AI legal: Concerns related to a scam, waste, and exploitation prevention

Roy et al. studied an important development in healthcare is the application of AI to the identification and mitigation of fraud, waste, and abuse (FWA) in payment systems. This kind of problems are being addressed by healthcare researchers using AI as presented in Fig. 10.5 showing the flow diagram of AI technology in clinical research. Businesses are using AI more and more to improve the efficacy and precision of FWA prevention, and some noteworthy success stories are starting to emerge. One such example presented in their study is a large health insurance that claimed to have saved $1 billion annually by implementing AI-driven FWA (Roy, 2022) preventive measures. This demonstrates the significant influence artificial intelligence (AI) may have in reducing monetary losses brought on by fraud in healthcare payment systems. Healthcare organizations have seen significant cost reductions as a result of AI's revolutionary ability to detect abnormalities, anomalies, and possible cases of FWA.

There are, nonetheless, certain legal and regulatory issues with the environment. A judgment earlier this year by a federal appeals court has complicated the employment of AI to provide Medicare Advantage and

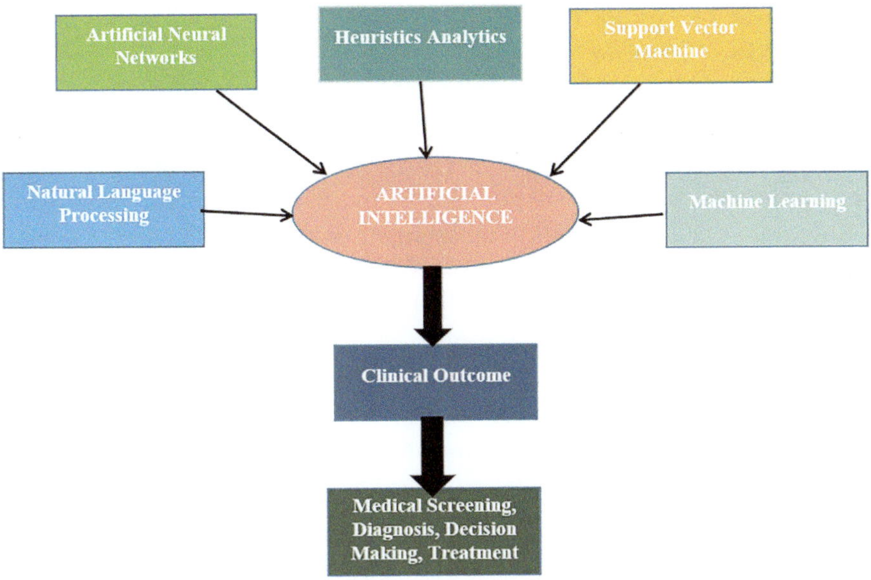

Figure 10.5 Flow process diagram of AI in healthcare.

Medicaid managed care organizations with prior authorization and usage management services. The court underlined the necessity of qualitatively reviewing AI applications, suggesting a degree of examination that may make the organization using AI accountable for its results.

The crucial nature of healthcare choices impacted by AI systems gives rise to the legal and regulatory ramifications. Liability and responsibility issues are brought up by the possibility of biases, mistakes, or unforeseen repercussions in AI systems. Regulators are closely monitoring the use of AI in healthcare, particularly in areas where it directly affects patient care and financial transactions, as the technology becomes increasingly integrated into the system.

Law and regulation frameworks must be carefully considered in order to ensure the morally and responsibly used AI in FWA prevention. It is critical to strike a balance between utilizing AI algorithms' efficiency and reducing possible hazards. Companies using artificial intelligence in healthcare payment systems need to manage the constantly changing regulatory environment to guarantee adherence and reduce legal risk.

Support Vector Machine

Heuristics Analytics

Artificial Neural Networks

Natural Language Processing
Machine Learning
Clinical Outcome
Medical Screening, Diagnosis, Decision Making, Treatment

Furthermore, another researcher Mackey et al. after careful consideration observed that there will probably be more legal discussion about AI in healthcare, which calls for the establishment of precise rules and regulations for AI applications in the prevention of FWA. It is critical that regulatory agencies, healthcare institutions, and AI developers work together to create comprehensive frameworks that protect against legal hazards and encourage the ongoing application of AI for effective FWA prevention. Even if artificial intelligence (AI) has proven to be highly effective in reducing instances of fraud, waste, and abuse in healthcare payment systems, the complexities of legal and regulatory issues are growing. In order to manage possible risks and guarantee the moral use of AI in healthcare, navigating this environment calls for a proactive strategy. Stakeholders must work together to create strong legal frameworks that support innovation and maintain patient care standards and financial integrity in healthcare transactions as the sector continues to adopt AI (Mackey et al., 2020).

Conclusion

In this paper, a detailed description have been provided across a range of different use cases of AI in Healthcare and the several Legal and Regulatory issues that are raised when artificial intelligence (AI) is incorporated into healthcare across a range of medical specialties. The complex legal environment, as demonstrated by individual cases, emphasizes the pressing need for all-encompassing and preventative actions. AI faces many difficulties when handling complex medical requirements. These difficulties highlight how important it is to collaborate across disciplines and build strong ethical AI standards.

The need to match AI developments with current legal frameworks is a common theme in a variety of healthcare scenarios as discussed earlier in use cases, such as reproductive health, HIV-related public health, juvenile justice, college mental health services, adolescent healthcare, breast cancer diagnosis, and fraud prevention. To ensure the proper implementation of AI technology, it becomes imperative to strike a balance between privacy restrictions, ethical concerns, and legal rights. The need for precise rules and regulations that strike a careful balance between personal preferences and prospective healthcare advantages is highlighted in some cases by the lack of conclusive court decisions.

In order to keep the proper use of AI it is essential that legal frameworks be modified and new laws to be made to reflect the rapidly changing AI technology landscape. This includes resolving privacy laws, reducing prejudice in AI systems, and making sure that legal frameworks are continuously updated to support moral behavior and safeguard patient privacy and life.

To guarantee responsible and ethical AI usage in healthcare, it becomes important to find the correct balance between innovation and protecting patient rights, especially in light of HIPAA and other changing rules. A thorough and cooperative strategy is necessary to address the complex legal issues surrounding artificial intelligence in healthcare. Establishing and modifying frameworks that encourage ethical and responsible usage of AI while negotiating the rapidly evolving landscape of healthcare technology is necessary and requires collaboration between stakeholders. In the rapidly changing field of artificial intelligence in healthcare, this strategy is crucial to preserving patient safety, maintaining accuracy in the distribution of medical information, and upholding the highest standards of care.

References

Awotunde, J.B., Jimoh, R.G., Folorunso, S.O., Adeniyi, E.A., Abiodun, K.M., Banjo, O.O., 2021. Privacy and security concerns in IoT-based healthcare systems. In: Internet of Things. Springer Science and Business Media Deutschland GmbH, Nigeria, pp. 105−134. https://doi.org/10.1007/978-3-030-75220-0_6, 21991081.

Barrington, N.M., Gupta, N., Musmar, B., Doyle, D., Panico, N., Godbole, N., Reardon, T., D'Amico, R.S., 2023. A bibliometric analysis of the rise of ChatGPT in medical research. Med. Sci. 11 (3), 61. https://doi.org/10.3390/medsci11030061, 2076−3271.

Bohr, A., Memarzadeh, K., 2020. The rise of artificial intelligence in healthcare applications. In: Artificial Intelligence in Healthcare. Elsevier, Denmark, pp. 25−60. https://doi.org/10.1016/B978-0-12-818438-7.00002-2, 9780128184387.

Carter, S.M., Rogers, W., Win, K.T., Frazer, H., Richards, B., Houssami, N., 2020. The ethical, legal and social implications of using artificial intelligence systems in breast cancer care. Breast 49, 25−32. https://doi.org/10.1016/j.breast.2019.10.001, 09609776.

Cath, C., 2018. Governing artificial intelligence: ethical, legal and technical opportunities and challenges. Phil. Trans. Math. Phys. Eng. Sci. 376 (2133), 20180080. https://doi.org/10.1098/rsta.2018.0080, 1364−503X.

Chaieb, S., Garrouch, K., Al-Ali, N.S., 2023. Perceptions of the use and benefits of artificial intelligence applications: survey study. J. Med. Artif. Intell. 6. https://doi.org/10.21037/jmai-23-59, 26172496, 28-28.

Faden, R.R., Geller, G., 1991. AIDS, Women, and the Next Generation: Towards a Morally Acceptable Public Policy for HIV Testing of Pregnant Women and Newborns. Oxford University Press.

Feld, A.D., 2005. The health insurance portability and accountability act (HIPAA): its broad effect on practice. Am. J. Gastroenterol. 100 (7), 1440–1443. https://doi.org/10.1111/j.1572-0241.2005.50621.x, 0002–9270.

Gerke, S., Minssen, T., Cohen, G., 2020. Ethical and legal challenges of artificial intelligence-driven healthcare. In: Artificial Intelligence in Healthcare. Elsevier, United States, pp. 295–336. https://doi.org/10.1016/B978-0-12-818438-7.00012-5, 9780128184387.

Glied, S.A., Frank, R.G., Wexler, J., 2019. Mental health disability, employment, and income support in the Twenty-First Century. In: The Palgrave Handbook of American Mental Health Policy. Springer Science and Business Media LLC, pp. 659–677. https://doi.org/10.1007/978-3-030-11908-9_23, 978-3-030-11907-2.

Harvey, H.B., Gowda, V., 2021. Regulatory issues and challenges to artificial intelligence adoption. Radiol. Clin. 59 (6), 1075–1083. https://doi.org/10.1016/j.rcl.2021.07.007, 00338389.

Houssami, N., Kirkpatrick-Jones, G., Noguchi, N., Lee, C.I., 2019. Artificial Intelligence (AI) for the early detection of breast cancer: a scoping review to assess AI's potential in breast screening practice. Expet Rev. Med. Dev. 16 (5), 351–362. https://doi.org/10.1080/17434440.2019.1610387, 1743–4440.

Khan, F., 2024. Regulating the revolution: a legal Roadmap to optimizing AI in healthcare. Minnes. J. Law, Sci. Technol. Spring.

Kirkpatrick, L., Collins, A., Harrison, E., Miller, E., Patterson, C., Sogawa, Y., Van Cott, A.C., Kazmerski, T.M., 2022. Pediatric neurologists' Perspectives on sexual and reproductive health care for adolescent and young adult women with epilepsy and intellectual disability. J. Child Neurol. 37 (1), 56–63. https://doi.org/10.1177/08830738211041824, 0883–0738.

Koocher, G.P., Kinscherff, R.T., 2016. Ethical Issues in Psychology and Juvenile Justice. American Psychological Association (APA), pp. 693–714. https://doi.org/10.1037/14643-031, 1-4338-1967-8.

Krittanawong, C., 2018. The rise of artificial intelligence and the uncertain future for physicians. Eur. J. Intern. Med. 48, e13–e14. https://doi.org/10.1016/j.ejim.2017.06.017, 09536205.

Mackey, T.K., Miyachi, K., Fung, D., Qian, S., Short, J., 2020. Combating health care fraud and abuse: conceptualization and prototyping study of a blockchain antifraud framework. J. Med. Internet Res. 22 (9), e18623. https://doi.org/10.2196/18623, 1438–8871.

Molnár-Gábor, F., 2019. Artificial intelligence in healthcare: doctors, patients and liabilities. In: Regulating Artificial Intelligence. Springer International Publishing, Germany, pp. 337–360. https://doi.org/10.1007/978-3-030-32361-5_15, 9783030323615.

Murdoch, B., 2021. Privacy and artificial intelligence: challenges for protecting health information in a new era. BMC Med. Ethics 22 (1), 1–5. https://doi.org/10.1186/s12910-021-00687-3, 1472–6939.

Naik, N., Hameed, B.M.Z., Shetty, D.K., Swain, D., Shah, M., Paul, R., Aggarwal, K., Ibrahim, S., Patil, V., Smriti, K., Shetty, S., Rai, B.P., Chlosta, P., Somani, B.K., 2022. Legal and ethical consideration in artificial intelligence in healthcare: who takes responsibility? Front. Surg. 9, 266. https://doi.org/10.3389/fsurg.2022.862322, 2296–875X.

Opderbeck, D.W., 2019. Artificial Intelligence in pharmaceuticals, biologics, and medical devices: present and future regulatory models. Fordham L. Rev. 88, 553.

Pesapane, F., Codari, M., Sardanelli, F., 2018. Artificial intelligence in medical imaging: threat or opportunity? Radiologists again at the forefront of innovation in medicine. Europ. Radiol. Experim. 2 (1). https://doi.org/10.1186/s41747-018-0061-6, 2509–9280.

Prakash, S., Balaji, J.N., Joshi, A., Surapaneni, K.M., 2022. Ethical conundrums in the application of artificial intelligence (AI) in healthcare—a scoping review of reviews. J. Personal. Med. 12 (11). https://doi.org/10.3390/jpm12111914, 2075–4426.

Raine, S.P., 2023. Legal and ethical issues in the Neurology of reproductive health. Neurol. Clin. 41 (3), 533–541. https://doi.org/10.1016/j.ncl.2023.03.003, 07338619.

Ramasamy, L.K., Khan, F., Shah, M., Prasad, B.V.V.S., Iwendi, C., Biamba, C., 2022. Secure smart wearable computing through artificial intelligence-enabled internet of things and cyber-physical systems for health monitoring. Sensors 22 (3), 1076. https://doi.org/10.3390/s22031076, 1424–8220.

Roy, S., 2022. Privacy prevention of health care data using AI. J. Data Acquis. Process. 37 (3), 769.

Schönberger, D., 2019. Artificial intelligence in healthcare: a critical analysis of the legal and ethical implications. Int. J. Law Info Technol. 27 (2), 171–203. https://doi.org/10.1093/ijlit/eaz004, 0967–0769.

Sharko, M., Jameson, R., Ancker, J.S., Krams, L., Webber, E.C., Rosenbloom, S.T., 2022. State-by-State variability in adolescent privacy laws. Pediatrics 149 (6), e2021053458. https://doi.org/10.1542/peds.2021-053458, 0031–4005.

Sisk, B.A., Canavera, K., Sharma, A., Baker, J.N., Johnson, L.-M., 2019. Ethical issues in the care of adolescent and young adult oncology patients. Pediatr. Blood Cancer 66 (5), e27608. https://doi.org/10.1002/pbc.27608, 1545–5009.

Tessmer-Tuck, J.A., Poku, J.K., Burkle, C.M., 2014. When courts intervene: public health, legal and ethical issues surrounding HIV, pregnant women, and newborn infants. Am. J. Obstet. Gynecol. 211 (5), 461–469. https://doi.org/10.1016/j.ajog.2014.04.014, 00029378.

Thomas, C.R., 2020. Ethical and legal issues in the care of transgender youth in the juvenile justice system. Ethics, Med. Public Health 13, 100464. https://doi.org/10.1016/j.jemep.2020.100464, 23525525.

Vyas, S., 2023. AI in Healthcare: Trends and Future Perspectives. CRC Press.

Wang, X., Sanders, H.M., Liu, Y., Seang, K., Tran, B.X., Atanasov, A.G., Qiu, Y., Tang, S., Car, J., Wang, Y.X., Wong, T.Y., Tham, Y.-C., Chung, K.C., 2023. ChatGPT: promise and challenges for deployment in low- and middle-income countries. Lancet Reg. Health - Western Pacific 41. https://doi.org/10.1016/j.lanwpc.2023.100905, 26666065.

WIS AI, 2018. Artificial Intelligence (AI) in Healthcare and Research. Nuffield Council on Bioethics, pp. 1–8.

Further reading

World Health Organization, 2021. Ethics and Governance of Artificial Intelligence for Health: WHO Guidance.

CHAPTER 11

Responsible and explainable artificial intelligence in healthcare: Conclusion and future directions

Akansha Singh[1], Krishna Kant Singh[2] and Ivan Izonin[3]
[1]School of CSET, Bennett University, Greater Noida, Uttar Pradesh, India; [2]Delhi Technical Campus, Greater Noida, Uttar Pradesh, India; [3]Lviv Polytechnic National University, Department of Artificial Intelligence, Lviv, Ukraine

Introduction

As we draw this investigation to a close, it is important that we reflect on the thread that we have pulled throughout this book, this important interplay of artificial intelligence and healthcare supported by a commitment to ethical responsibility and transparency. The chapters following this have been focused on building a detailed foundation in and around the multiple facets of AI in supporting health care delivery, but at the same time, ensuring that it is done in a fundamentally ethical and explainable manner. This chapter will thus serve to synthesize these discussions by giving a broad overview of the landscape of current responsible and explainable AI in healthcare, calling out a few challenges that currently exist and using that to open a visionary gaze toward future possibilities. It is this opportunity for us to distill the essence of all the insights that our conversation has yielded and make space for the unleashing of the next waves of innovations that continue to transform the landscapes of healthcare globally (Goyal et al., 2023) (Fig. 11.1).

Ethical foundations

Earlier discussions have ethically navigated the landscape that underlies AI use in healthcare. Lying at the heart of these discussions are the key ethical principles of fairness, transparency, and accountability—principles that form

Responsible and Explainable Artificial Intelligence in Healthcare
ISBN 978-0-443-24788-0
https://doi.org/10.1016/B978-0-443-24788-0.00011-X

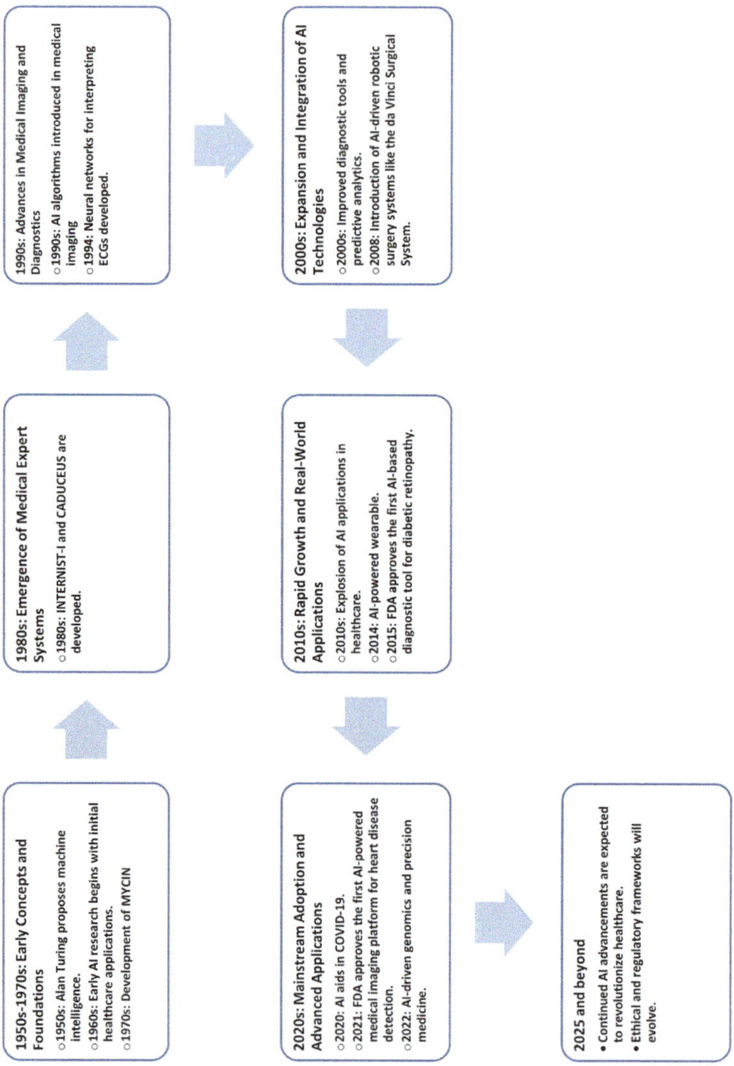

Figure 11.1 AI impact timeline for healthcare.

the pivot in the development of AI technologies and their implementation within sensitive and high-stake environments, such as healthcare.

Fairness has also been a recurrent theme, dealt with at quite some length in our analysis, with measures toward ensuring the same in algorithmic biases. There is a need to assure that the AI-created system will not perpetuate already existing disparities or will not even form new types of discrimination. Here, in this aspect, lies the importance of techniques such as bias detection and mitigation strategies that have to be evidenced so that AI can be customized to deliver with equity to patient demographics (Reith et al., 2022).

Another key principle has been transparency. AI is black box, and many of its systems are, therefore, challenging—more so in a field where the stakes involved concern human lives. The methodologies of explainable AI that make the workings of AI systems more understandable to users and stakeholders are being explored. Such transparency is vital for trust and further acceptance of AI technologies in healthcare.

Accountability entails clear lines of responsibility for the outcome of an AI intervention. We discussed the need to preserve high standards within regulatory frameworks to ensure the safe and responsible use of AI systems. Clear accountability will establish a system in which providers of healthcare and their patients can trust that the AI will serve as a tool for the improvement of healthcare outcomes in an ethical manner. Indeed, in this book, such principles were not only forming the theoretical framework but further exemplified through practical examples and case studies that really bridge the gap between abstract ethical considerations and real-world applications. This chapter builds upon these foundational ethics as we ponder the future trajectory of AI in healthcare.

Case studies and applications

AI has now become an indispensable tool in the healthcare sector, from experimental to improve considerably patient care and operational efficiency. We have selected a few illustrative case studies in this module to show successful integration of explainable AI in real-world healthcare practice settings. Each of them discusses the real-time advancements and practical benefits that AI-based solutions have brought about, but at the same time, it is also of substantial importance to facilitate explainability for purposes of gaining trust and clarity for clinicians and patients (Chaddad et al., 2023). In the process, this paper discusses case studies on how transparent AI processes assist in better decision-making, improved outcomes, and further cementing the symbiotic relationship between technology and human expertise in medicine.

1. **Imaging**

   Case Study: AI in the interpretation of diagnostic images, such as MRIs and CT scans, has shown great promise. Notable among these is the transparency with which this developed system provides information on its diagnostic process, hence making radiologists understand the rationality of some of its annotations and probability scores on the AI behind specific recommendations. This case enhanced the radiologists' diagnostic accuracy and increased their confidence in AI-assisted diagnostics.

2. **Individualized Treatment Plans**

   Case Study: In the field of oncology, several AI systems have been developed that are explainable and allow treatment plans to be personalized based on the genetic profile of individual patients. A pioneering study demonstrated an AI system capable of correlating vast datasets of genetic information with drug efficacy to suggest treatments and explain the recommendations in a manner such that oncologists could evaluate and apply effectively (Kiseleva et al., 2023).

3. **Patient Monitoring using Predictive Analytics**

   Case Study: An area in which the application of AI has significantly advanced predictive health care is the ability for patients to be monitored. In an application within a cardiac unit, AI was used to predict adverse events such as heart failure, using real-time data derived from patient monitors. It explained its predictions based on identified data patterns and anomalies, thereby allowing health professionals to take timely interventions.

4. **Management of Chronic Diseases**

   Case Study: One of the most original projects of this type was the development of an AI system to manage type 2 diabetes. The system applies machine learning in the analysis of patient history, impacts of his or her lifestyle, and current symptoms to predict levels of blood sugar and recommend insulin doses. What made this system unique was its ability to predict and explain predictions through several data touchpoints: dietary intake and physical activity, to give an insight into how lifestyle changes could have an effect on the condition of a patient, such that it could be understood both by patients and doctors (Goyal et al., 2023).

5. **Early Detection of Sepsis**
   Case Study: Sepsis is one of those disease states that result from infection and that, if left untreated, can become life-threatening in response to the infection. This hospital has implemented an AI program that is able to track patients' vital signs and laboratory results for early identification of sepsis. An AI system reports real-time alerts to medical staff with complete explanations of identified risk factors, which increase interventions in timely and precise treatment.
6. **Strengthening Mental Health Services**
   Case Study: Explainable AI methodologies have been utilized in predicting future episodes of depression and anxiety. Noteworthy among these is a therapeutic chatbot that modifies its interactions based on an AI model's interpretation of user sentiment and linguistic cues. In addition, the system explains to mental health professionals the basis for its therapy recommendations, giving insight into the emotional state of the patient and allowing better-informed decisions on therapies.
7. **Optimization of Hospital Operations**
   Case Study: A large healthcare system implemented a tool based on AI for optimal bed management in hospitals. Patient discharge predictions were made on the basis of data from many different departments, and the system recommended bed allocation to minimize waiting time and increase patient flow (Singh et al., 2022). Importantly, the AI was giving explanations of its advice, so the logic of the AI's operational suggestions could be understood by hospital administrators in order to foster greater acceptance and implementation efficiency.

   These detailed case studies illustrate how flexible and indispensable the value of explainable AI is in a range of healthcare fields. These applications are instances not just of the development of many aspects of health care but also of the creation of pivotal trust between AI systems and their human users: the way of gaining more widespread acceptance and integration in health care.

Technological advancements

Indeed, it is technological advancements that have taken place nowadays, especially in the use of deep learning and neural networks in the sector of artificial intelligence, which have altogether transformed the landscape of healthcare. These two technologies have played a critical role in shaping modern diagnostics, treatment planning, and the management of patients, ushering in a new phase of precision medicine and efficient delivery of health services.

Deep learning is a specialized form of machine learning that uses layers of neural networks to analyze large datasets. Their application in diagnostics has been more than anywhere else. A huge number of medical images are used to train it in order for the model to recognize complex patterns and anomalies with high accuracy. For example, they have been applied in detecting the early signs of a disease such as cancer and diabetic retinopathy in more cases than not, identifying rather faint markers that a human may have overlooked (Goyal et al., 2023). That, of course, saves time in diagnosis and, in most cases, ensures accuracy, which can be crucial for early intervention and, therefore, saving more lives.

Even more, neural networks are what has made it possible to change treatments, especially in such complex fields as oncology. Such AI systems are those that can combine heterogeneous types of data within the framework of the electronic medical records, genetics, and imaging studies in order to be able to develop a personalized regimen for different treatment methods. Due to the special features regarding the pathological state of the patient, AI can predict the ways in which he or she will react to these or those treatments, thereby aiding the clinician in making choices regarding the most effective treatments while suffering the least negative consequences. Such an approach makes up the basis of the precision medicine: treatment that is specifically individualized toward a given patient. Besides diagnostics and treatment, AI sees huge usage in the field of patient management. The application of machine learning models in the real-time monitoring of patient health is increasing daily, allowing for predictions of potential health deteriorations and management of chronic conditions in a more efficient manner. In ICUs, AI-driven systems analyze real-time data from different monitors to predict acute events such as sepsis or cardiac arrests to avoid that fatal moment. In patients with chronic diseases, AI models help adjust medications and suggest lifestyle changes based on continuous data input from wearable devices and patient-reported outcomes. Such a proactive management approach facilitates the maintenance of stable health conditions and the avoidance of rehospitalization.

Further, AI is integrated with other emergent technologies to magnify benefits across the health system. For example, the coupling of AI with the Internet of Things provides better data on health through connected devices. Robotics systems in surgical settings, augmented with AI, help the surgeon provide levels of precision that result in better outcomes and shorter recovery times. Furthermore, blockchain is increasingly used with AI to ensure the security of humongous data generated by the technologies to ensure patient privacy and trust.

And that's only the beginning with these technological breakthroughs. With the constant evolution of AI, quite literally every aspect of delivery and management within the healthcare system is going to transform. This will not only enhance the capabilities of healthcare professionals, but also bring about improvements in enormous patient care and health system efficiencies. These continuations in the development and integration of AI within healthcare create visions for a future where medical care is going to be so much more accurate, personalized, and accessible than it has ever been.

Major challenges and resolutions

Data privacy and security

Data security and privacy remain some of the biggest challenges as AI technologies are integrated into an increasingly connected healthcare system. This is because the acquisition, storage, and processing of big data are very much needed for training and refinement purposes in AI models, and along with it are highly risky in terms of privacy, especially sensitive medical data. In addition, the patient consent issue hosts a number of ethical predicaments that could be further complicated by the complex data-sharing systems characteristic of modern digital healthcare. Of great importance is the protection of the data from breaches, unauthorized access, or misuse. Additionally, AI systems will detect unexpected insights and might, therefore, be difficult to inform when seeking consent for data collection. Patients should be informed on the implications, risks, and measures of security taken with their privacy. This is where differential privacy and other solutions could solve the problems, making it possible for AI system learning capabilities and researchers to learn from data without incurring privacy effects on individual patients in a data set (Jangra et al., 2023). Differential privacy is achieved by the addition of a small amount of noise in a data set so that individual contributions get masked, but without much distortion of the insight that can be drawn with AI analysis. Another innovative solution was federated learning, in which the AI algorithm itself is taken to the location of the data, trained on the data locally, and then the learned information is collected to improve the model. It has the reduction of moving sensitive information and reduces the risk of moving information, minimizing the risks of a data breach.

Bias and fairness

The presence of algorithmic bias in AI applications raises gigantic ethical concerns, particularly in healthcare, which is highly sensitive. Sources of such bias may come from skewed datasets in training, flaws in the design of the algorithm, or interpretation of AI outputs. Biased AI models can even widen health disparities and have greater impacts on marginalized populations. For example, if a dataset is dominated by some demographic groups, then AI developed using this dataset is likely to perform poorly for under-represented groups (Singh et al., 2021). Solutions require the diversity and representativeness of the datasets: data must be collected from the whole range of populations so that the AI systems treat people fairly. Another key element is ongoing monitoring and audit of AI systems to remove bias. Tools and methodologies to evaluate AI decisions for fairness contribute to the discovery of bias and its correction (Allen, 2024). Some common approaches to algorithmic fairness are re-weighting training data and adjusting the algorithmic thresholds of decision to ensure fairness to under-represented groups.

Regulatory frameworks

The rapid development of AI in healthcare regulation has significant impacts on the ways that AI solutions are deployed and scaled. Key among the challenges is the diversity of regulatory standards among regions and countries, which therefore creates barriers to international deployment and complicates compliance for multinational health care entities. Add to these rapid developments in technology, which often overtake processes in the regulatory environment; potential gaps in oversight can emerge. There are global initiatives to develop more harmonized international standards and regulations in the field of health and AI (Alves et al., 2023). The WHO and the International Medical Device Regulators Forum have initiatives in this area at the international level, to harmonize standards and guidelines globally in a manner that allows innovation to be balanced against safety and efficacy. Many national regulatory agencies, such as the US Food and Drug Administration, are in the process of updating or creating entirely new pathways to more efficiently manage unique issues raised by AI technologies—for instance, creating a precertification program for innovations in digital health. That way, solving these challenges by advanced technological solutions and thoughtful regulatory approaches will usher in the future of AI in healthcare, efficacious and ethically responsible in such a way as to ensure patient care and operational efficiency without any compromise in high ethical standards.

Future directions

The future for health-related AI promises to be revolutionary—facilitated through rapidly growing technological advances, interdisciplinary collaboration, and dedication to patient-centered care. These developments promise to bring not just more effective delivery of healthcare but will also ensure that it is more personalized, inclusive, and ethically grounded.

Technological advances in AI

In fact, the AI-technology era at the vanguard is flowering with stunning innovations that have the potential to literally redesign the healthcare systems. Probably the most outstanding of them is quantum computing. In its infancy, quantum computing can be the solution to processing complex datasets at speeds many orders of magnitude faster than classical computers. It will change the way to model biological processes, simulate the effect of drugs, or decode genetic information. Some examples of this are efficient analyses of large databases of molecular structures in discovery phases of new pharmaceuticals via quantum algorithms and the ability of quantum computing to predict their interactions much faster than it is currently possible (Jiang et al., 2017).

Another major step has been in the development of advanced neural network architectures. Techniques such as transformer neural networks, which have revolutionized natural language processing, are now making inroads into healthcare. The mentioned architectures are especially well fit to manage sequenced data and are therefore ideal for applications such as the analysis of the patient health records for the prediction of patient outcomes based on historical data. Recently, AI advancements through reinforcement learning have focused on AI systems learning optimal actions through trial and error and are beginning to be applied in personalized medicine and robotic surgery so that such systems can adapt and learn optimal strategies in real time.

Interdisciplinary collaboration

The complexity of modern health care problems and challenges does not allow a single discipline to address them effectively. Thus, interdisciplinary collaboration is extremely important for computer scientists, ethicists, health care providers, and policymakers in the development of AI systems that are technologically advanced and at the same time ethically sound and socially beneficial (Mondal et al., 2020). While computer scientists do all the work in the actual development of algorithms and data analytics, ethicists ensure that the development and use of these technologies are done in a manner that respects patient autonomy, privacy, and fairness. Healthcare providers further bring insight into the practical requirements and limitations of working healthcare environments, ensuring that AI-driven solutions are clinically relevant and truly contribute to patient care. However, policymakers provide the last assurance of support for the frameworks that protect public health and do so in a way that fosters innovation (Johnson et al., 2018). Such collaborations could provide a framework for guidelines and standards development in health AI and foster best practices for its development with safety, efficacy, and ethical integrity. Such interdisciplinary teams will further push public engagement and trust in AI technologies, allowing a wider understanding and acceptance of their benefits.

Patient-centered approaches

Patient-centered AI systems in health care are moving toward being more patient-centric. This approach involves the tailoring of health services to the needs and preferences of the patient for a specific health context. AI technologies are at the vanguard of such moves, having developed tools for personalized medicine, patient engagement, and health management.

AI-powered personalized treatment plans analyze individual patient data ranging from genetic profiles to lifestyle choices in order to tailor treatments that are most likely to be effective for specific patients. This not only improves outcomes but also minimizes the risk of side effects (Jiang et al., 2017).

Another sector where AI is really impacting is patient engagement. AI-powered applications can provide patients with updates on their health statuses at regular intervals, reminders to take medications, and recommendations on health measures. This way, the providers and the patients are enabled to maintain a continuous interaction that goes beyond the conventional clinical settings.

AI can also be used to predict patient preferences and improve the experiences within healthcare facilities. From appointment schedule optimization and the reduction of waiting time to even predicting the flow of patients, AI helps in enhancing health services to be more responsive to patients. The future of AI in healthcare does not lie in the countless technological advances but rather in making the healthcare system efficient, effective, and, most importantly, patient-centered. It will progress with a technological integration approach that fosters broad-based collaboration and focus on patient-centeredness in the approaches that guide the evolution of healthcare, with personalization and access never seen before.

Conclusion: Shaping the future of AI in healthcare

As we round out our responsible and explainable artificial intelligence in healthcare discussion, it is important for everybody to make a mind shift—from technologists and clinicians to policymakers and patients. Getting AI embedded into health is not just about leveraging technology; it is about doing so in an ethical, transparent, and just manner. This calls for a deliberate effort to entrench and promote standards that respect the dignity and rights of all individuals.

Call to action

To these many stakeholders engaged across the landscape of health care AI, therefore, the clarion call hence forth is this: Commit to developing and deploying AI technologies that work effectively, are fair, and remain interpretable. Healthcare providers should expect that AI tools would be able to provide clear and concise reasoning for any recommendations provided in such a manner as to augment clinical intuition and expertise, rather than obfuscate those (Sharma et al., 2023). Technologists and developers must further their innovation in a way that emphasizes both accuracy and responsible, ethical design and deployment. Policymakers and regulators must develop and enforce regulations that keep AI standards in health care high, protecting patient privacy while maintaining fairness in AI-generated outcomes.

Vision for the future

The potential for AI to revolutionize health is massive and clear. Picture a world where AI increases the effectiveness of health systems, enhances the predictability of care, and lays out the principles of personalization. In this future, diagnosis is made much more accurate, quick enough to lead to the early detection of problems and timely intervention. The treatment package is designed for the person not only on a medical background but based on their genetic data, way of life, and personal choice.

More so, it will make healthcare democratic: advanced diagnostics and personalized treatment plans that can be fueled beyond high-resource regions and into areas that were, until hitherto, undeserving in quality, will bring down disparities in health care delivery. AI also has the potential to be a game-changer in managing crises, for example pandemics, where the use of AI can speed up responses and optimize resource allocation.

But to bring this future to light, AI development and operationalization must be done with unshakeable commitments to fairness, accountability, and transparency. This realization brings the global healthcare community together to have AI systems that are not just technically accurate but are culturally sensitive and socially inclusive. Then and only then could AI fulfill its potential as an enabling force for good within healthcare.

This is a need, not a goal. As we stand on the verge of that technological revolution, do everything in your capability to push AI toward a future that enhances every aspect of healthcare—predictive, preventive, personalized, and participatory. May this be our shared vision and commitment as we move forward, shaping an AI-enhanced healthcare landscape characterized by humanity as much as innovation.

References

Allen, B., 2024. The promise of explainable AI in digital health for precision medicine: a systematic review. J. Personal. Med. 14 (3). https://doi.org/10.3390/jpm14030277, 2075-4426.

Alves, P., Hassanzadeh Benam, A., Hariri, F., 2023. Artificial intelligence ethics and challenges in healthcare applications: a comprehensive review in the context of the European GDPR mandate. In: Machines Learning Knowledge Extraction. https://doi.org/10.3390/make5030053.

Chaddad, A., Peng, J., Xu, J., Bouridane, A., 2023. Survey of explainable AI techniques in healthcare. Sensors 23 (2). https://doi.org/10.3390/s23020634, 1424-8220.

Goyal, L., Dhull, A., Singh, A., Kukreja, S., Singh, K.K., 2023. VGG-COVIDNet: a Novel model for COVID detection from X-Ray and CT Scan images. Proc. Comput. Sci. 218, 1926—1935. https://doi.org/10.1016/j.procs.2023.01.169, 18770509.

Jangra, M., Dhull, S.K., Singh, K.K., Singh, A., Cheng, X., 2023. O-WCNN: an optimized integration of spatial and spectral feature map for arrhythmia classification. Complex Intell. Syst. 9 (3), 2685—2698.

Jiang, F., et al., 2017. Artificial intelligence in healthcare: past, present and future. Stroke Vascul. Neurol. https://doi.org/10.1136/svn-2017-000101.

Johnson, K.W., et al., 2018. Artificial intelligence in cardiology. J. Am. Coll. Cardiol. https://doi.org/10.1016/j.jacc.2018.09.068.

Kiseleva, et al., 2023. Editorial: explainable artificial intelligence for critical healthcare applications. Front. Digit. Health. https://www.frontiersin.org/articles/10.3389/fdgth.2023.00007/full.

Mondal, S.S., Mandal, N., Singh, A., Singh, K.K., 2020. Blood vessel detection from retinal fundas images using GIFKCN classifier. Proc. Comput. Sci. 167, 2060–2069. https://doi.org/10.1016/j.procs.2020.03.246, 18770509.

Reith, D., et al., 2022. Responsible AI in healthcare: opportunities, challenges, and best practices. Front. Comput. Sci.

Sharma, U., Ganapathi, A., Singh, A., Singh, K.K., 2023. Blockchain in healthcare: use cases. Blockchain and Deep Learning for Smart Healthcare 147–169.

Singh, A., Singh, K.K., Greguš, M., Izonin, I., 2022. CNGOD-an improved convolution neural network with grasshopper optimization for detection of COVID-19. Math. Biosci. Eng. 9, 12518–12531.

Singh, K.K., Singh, A., Lin, J.W., Elngar, 2021. Deep Learning and IoT in Healthcare Systems: Paradigms and Applications. CRC Press, 2021.

Further reading

Nature, 2023. Explainable artificial intelligence (XAI) in healthcare: current status and future directions. Nat. Med. https://www.nature.com/articles/s41746-023-00751-9.pdf.

Index

'*Note:* Page numbers followed by "f" indicate figures and "t" indicate tables.'

A

Accessibility, 147
Accountability, 95–100, 103–104, 144–145
Adherence applications, 136–137
Adolescent patient, 272–273
Algorithm design, 123
Algorithm 1 implementation results, 240–245, 245t
Algorithm 2 implementation results, 245–246, 245t–246t
Algorithm 3 implementation results, 246–247, 247t–248t
Ambient assisted living, 172–180
  assistive robots, 177–179
  cognitive assistants, 179
  smart home, 172–176
  social/emotional stimulation, 179–180
AR. *See* Augmented reality (AR)
Artificial intelligence (AI)
  accountability, 95–100, 103–104
  age, 61
  AI-aided drug discovery efforts, 98–99
  ambient assisted living, 172–180
    assistive robots, 177–179
    cognitive assistants, 179
    smart home, 172–176
    social and emotional stimulation, 179–180
  biomedical data sharing, 101–102
  black box, 97–98
  cardiac arrest prediction tools, 97
  challenges, 148–151
    data bias, 148–149
    data privacy, 150
    fear of change, 150
    high cost, 150–151
    limited data, 148
    security, 150
    standardization, 149
    transparency, 149
    understanding, 149–150
  challenges of, 94–95
  chest pain type, 61
  cholesterol, 62
  classification task description, 60–65
  clinical practice, 99
  correlation between features, 71–72, 72f–73f
  data analysis, 60–72
  data sample, 65
  data set features, 65
  data visualization, 60–72, 66f–68f
  decision boundary, 83
  decision tree machine learning model, 84
  designing transparent frameworks, 100–104
  diagnosis and treatment
    accessibility, 147
    accountability, 144–145
    autonomy, 142
    collaboration, 147–148
    data privacy, 143–144
    data quality enhancement, 146–147
    equity, 147
    ethical principles in, 140–148, 141f
    fairness principles, 145–146
    inclusivity, 147
    liability, 144–145
    nondiscrimination, 145–146
    risk reduction and safety, 142–143
    trustworthiness, 146
    validity, 146
  electrocardiogram (ECG), 63–64, 63f
  ethical and responsible practices framework, 102
  exercise angina, 64
  explainability, 96
  exploring, 92

Artificial intelligence (AI) (*Continued*)
fasting blood sugar, 62
future of, 151
healthcare, 128–135, 157–158
adherence applications, 136–137
administrative applications, 138–140
diagnosis, 137–138
expert system, 134
machine learning (ML), 130–132, 132f
natural language processing (NLP), 132–133
patient engagement, 136–137
robots and robotic process, 134–135
treatment applications, 137–138
healthcare trustworthiness methods, 58–59, 60f
machine learning (ML), 92–93, 93f
Max HR, 64
medical settings, 94
medical visualization, 158–169
augmented reality (AR), 159
computer vision (CV), 159
health monitoring and wearables, 161–165
intelligent personal health records, 160–161
natural language processing, 165–167
patient experience, 160
personal records integration, 167–169
surgery and diagnosis, 159
virtual reality (VR), 159
model comparison, 78–84
model creation, 74–77, 74f
model types, 73–74, 74t
navigating, 95–100
neural network, 77–78, 77f
oldpeak, 64, 69f
permutation feature importance, 80–82
privacy framework, 101–102
random forest confusion matrix, 76f
recursive feature selection, 82–83
resting BP, 61–62
SHAP, 80–82
smart devices, 169–172
minimally invasive surgery, 170
neuroprosthetics, 170–172
ST slope, 64–65
target class, 71, 71f
transforming healthcare, 99
transparency, 93, 94f, 95–100
framework, 103–104
uses of, 99–100
Artificial neural networks, 4
Assistive robots, 177–179
Augmented reality (AR), 159

B

Beneficence, 40–41
Bias, 31–34, 292
advanced algorithmic design, 119
algorithmic bias, 115
assessment techniques, 118
bias detection, 118
continuous improvement and adaptation, 119
data annotation, 116
data bias, 114–115
data collection, 115–116
deployment and feedback loops, 117
diverse collection, 118
identifying sources of, 115–117
impact of, 117–118
label bias, 115–116
mitigating bias strategies, 118–119
model and algorithmic bias, 116
regulatory and ethical frameworks, 119
representative data, 118
sources of, 114–119
stakeholder education and involvement, 119
types of, 113–114
Breast cancer, 273–274

C

Calibration, 121–122
Case studies/applications, 287–289
Chest pain type, 61
Cholesterol, 62
Chronic diseases, 288
Classification task description, 60–65
Cognitive assistants, 179
College mental health, 270–272

Computer vision (CV), 5, 159
Confidentiality, 37
Connected/augmented care, 9–10
  AI-powered chatbots, 10
  ambient and intelligent care, 10
  digital personal assistants, 10
Control, 35
Cybersecurity, 38–39

D

Data
  analysis, 60–72
  balancing results, 240, 241t–242t
  collection, 123
  privacy, 36–37, 143–144, 291–292
  processing, 123
  quality enhancement, 146–147
  sample, 65
  security, 291–292
Data set features, 65
Dataset 1 proposed algorithm, 201–203
Dataset 2 proposed algorithm, 203
Datasets description, 235–239
  dataset 1, 235–237
  dataset 2, 237
  dataset 3, 237–238
  small data analysis, balancing techniques for, 238–239
Data visualization, 60–72, 66f–68f
Decision boundary, 83
Decision tree machine learning model, 84
Deep learning, 6
Deployment monitoring, 123
Designing transparent frameworks, 100–104
Discrimination, 34
Disparate impact, 121
Doctor ChatGPT, 275–277

E

Electrocardiogram (ECG), 63–64, 63f
Emotional stimulation, 179–180
Equality, 34
  opportunity, 121
Equity, 147
Ethical considerations, 285–287
  autonomy, 34–35
  beneficence, 40–41
  bias, 31–34
  confidentiality, 37
  control, 35
  cybersecurity, 38–39
  data privacy, 36–37
  discrimination, 34
  equality, 34
  explainability, 38
  fairness, 34
  freedom, 34–35
  health domain, AI applications in, 45–48
  informed consent, 36
  justice, 29–31
  methodology, 28–29, 30t, 31f–33f
  responsibility, 41
  solving strategies, 43–45, 44t–46t
  supplementary ethical issue, 41–42
    solidarity, 41–42
    sustainability, 42
  transparency, 37–38
  trust, 39–40
Ethical/responsible practices framework, 102
Ethical/social nature, 20
Exercise angina, 64
Expert systems, 4, 134

F

Fairness, 34, 292
  challenge of bias, 110–112
  clinical decision making, 108–109
  cost-efficiency, 109
  distributive fairness, 121
  implementing, 122–123
  importance of, 112–113
  interactional fairness, 121
  measurement of, 121–122
  patient care enhancements, 109–110
  personalized medicine, 108
  predictive diagnostics, 108
  principles, 145–146
  procedural fairness, 120
  reduction, 109
  remote monitoring, 109

Fairness (*Continued*)
representational fairness, 121
risks and consequences, 110–112
telemedicine, 109
Fasting blood sugar, 62
Feedback loop, 123
Freedom, 34–35
Future vision, 295–296
Fuzzy logic, 4

H

Healthcare, 6–13
artificial intelligence today, 8
artificial neural networks, 4
computer vision, 5
connected/augmented care, 9–10
AI-powered chatbots, 10
ambient and intelligent care, 10
digital personal assistants, 10
deep learning, 6
definition, 2–6
ethical and social nature, 20
experts requirement, 17–18
expert systems, 4
fuzzy logic, 4
imbalance, 19–20
international trends, 13–14
issues and challenges, 18–19
long-term artificial intelligence, 8–13
machine learning, 4–5
medical procedures stages
clinical studies, 15
diagnostic biomarkers, 16
possibilities eliminating, 15
problematic region locating, 15
medium-term artificial intelligence, 8
natural language processing, 5
positive developments, 19–20
potential and advantages in, 16
precision diagnostics, 10–12
diabetic retinopathy screening, 11
diagnostic imaging, 10–11
radiotherapy planning, precision and reducing waiting timings for, 11–12
precision medicine
AI-empowered healthcare professionals, 13
new curative therapies, 12–13
precision therapeutics, 12
AI-driven drug discovery, 12
immunomics and synthetic biology, 12
reasonableness, 19–20
robotics, 5–6
Healthcare trustworthiness methods, 58–59, 60f
Health domain, AI applications in, 45–48
Health monitoring and wearables, 161–165
HIV-related public health, 268–269
Hospital operations optimization, 289
Hybrid cascade-based prediction model, 229–235
first algorithmic implementation proposed method, 231
improved performance and interpretability design, 229–230
second algorithmic implementation proposed method, 231–233
third algorithmic implementation proposed method, 233–235
Hybrid ensemble learning model
algorithm 1 implementation results, 240–245, 245t
algorithm 2 implementation results, 245–246, 245t–246t
algorithm 3 implementation results, 246–247, 247t–248t
data balancing results, 240, 241t–242t
datasets description, 235–239
dataset 1, 235–237
dataset 2, 237
dataset 3, 237–238
small data analysis, balancing techniques for, 238–239
hybrid cascade-based prediction model, 229–235
first algorithmic implementation proposed method, 231
improved performance and interpretability design, 229–230

second algorithmic implementation proposed method, 231–233
third algorithmic implementation proposed method, 233–235
interpreted artificial intelligence (IAI), 222
investigated methods, 249–250, 249f–250f
interpretability, 253–255
training duration, 251–253, 251f–252f
Ito decomposition, 227–228
linear polynomial synthesis, 228–229
Naïve Bayes classifier, 223–224
results modeling, 239–247
successive geometric transformations model (SGTM), 224–227, 225f, 227f

I

Imbalance, 19–20
Inclusivity, 147
Individualized treatment plans, 288
Informed consent, 36
Intelligent personal health records, 160–161
Interdisciplinary collaboration, 294
Interpreted artificial intelligence (IAI), 222
IOT technology and wearables, 277–278
Ito decomposition, 227–228

J

Justice, 29–31
Juvenile justice, 269–270

L

Legal/regulatory issues
AI in healthcare
adolescent patient, 272–273
benefits, 262
breast cancer, 273–274
challenges, 262
college mental health, 270–272
Doctor ChatGPT, 275–277
gift/threat, 263–266
HIV-related public health, 268–269
juvenile justice, 269–270
legal and ethical concerns, 267–268
mothers and new-borns, 268–269
personal information safeguarding patients, 274–275
reproductive health, neurology, 267–268
healthcare market trends, 260–261
IOT technology and wearables, 277–278
landscape, 261–262
legal and ethical problems raised, 261
legal and regulatory matters, 261–262
scam, waste, and exploitation prevention, 278–280
Liability, 144–145
Linear polynomial synthesis, 228–229

M

Machine learning (ML), 4–5, 92–93, 93f, 130–132, 132f
Max HR, 64
Medical image classification, deep learning approach for
evaluation of models, 199–200
materials and methods, 196–198
modeling and results, 200–203
analysis tools, 200–201
applying results, 201–203
dataset 1 proposed algorithm, 201–203
dataset 2 proposed algorithm, 203
model development, 200–201
proposed model architecture, 196–198
selected datasets analysis, 191–196
state-of-the-arts, 185–191
training, 199–200
validation, 199–200
XAI methods, 203–215
dataset 1, 205–212, 208f, 210f–212f
dataset 2, 212–215, 213f–214f, 216f
Medical procedures stages
clinical studies, 15
diagnostic biomarkers, 16
possibilities eliminating, 15
problematic region locating, 15
Medical settings, 94
Methodology, 28–29, 30t, 31f–33f
Minimally invasive surgery, 170
ML. *See* Machine learning (ML)
Model comparison, 78–84

Model creation, 74–77, 74f
Model types, 73–74, 74t

N

Naïve Bayes classifier, 223–224
National Eating Disorder Association (NEDA), 275–276
Natural language processing (NLP), 5, 132–133, 165–167
Neural network, 77–78, 77f
Neuroprosthetics, 170–172
Nondiscrimination, 145–146

O

Oldpeak, 64, 69f

P

Patient-centered approaches, 294–295
Patient engagement, 136–137
Patient monitoring, 288
Permutation feature importance, 80–82
Personal information safeguarding patients, 274–275
Personal records integration, 167–169
Positive developments, 19–20
Precision diagnostics, 10–12
  diabetic retinopathy screening, 11
  diagnostic imaging, 10–11
  radiotherapy planning, precision and reducing waiting timings for, 11–12
Precision medicine
  AI-empowered healthcare professionals, 13
  new curative therapies, 12–13
Precision therapeutics, 12
  AI-driven drug discovery, 12
  immunomics and synthetic biology, 12
Predictive analytics, 288
Privacy framework, 101–102
Proposed model architecture, 196–198

R

Random forest confusion matrix, 76f
Reasonableness, 19–20
Recursive feature selection, 82–83
Regulatory frameworks, 292–293
Reproductive health, neurology, 267–268
Responsibility, 41
Resting BP, 61–62
Risk reduction, 142–143
Robotics, 5–6, 134–135

S

Safety, 142–143
Scam, waste, and exploitation prevention, 278–280
Selected datasets analysis, 191–196
Sepsis, 289
SHAP, 80–82
Smart devices, 169–172
  minimally invasive surgery, 170
  neuroprosthetics, 170–172
Smart home, 172–176
Social stimulation, 179–180
Solidarity, 41–42
Strengthening mental health services, 289
ST slope, 64–65
Successive geometric transformations model (SGTM), 224–227, 225f, 227f
Sustainability, 42

T

Target class, 71, 71f
Technological advancements, 157, 289–291
Technological advances, 293
Testing, 123
Transparency, 37–38, 93, 94f, 95–100
  framework, 103–104
Trust, 39–40
Trustworthiness, 146

V

Validation, 123
Validity, 146
Virtual reality (VR), 159

X

XAI methods, 203–215
  dataset 1, 205–212, 208f, 210f–212f
  dataset 2, 212–215, 213f–214f, 216f